CIA IMPROVISED SABOTAGE DEVICES

Desert Publications
El Dorado, AR 71730 U. S. A.

CIA
Improvised Sabotage Devices

215 S.Washington Ave.
El Dorado,AR 71730 U.S.A.
870-862-3811
www.deltapress.com

ISBN 978-0-87947-209-2
12 11 10 9 8
Printed in U.S.A.

The Publisher Desert Publications is a division of
The DELTA GROUP, Ltd.
Direct all inquiries & orders to the above address.

CONTENTS

ILLUSTRATIONS

Page

INTRODUCTION

For quite a few years now, considerable amounts of time, effort, and money have been expended on improving older sabotage devices and accessory gear and in developing new and better items. As a result, a wide variety of manufactured explosive and incendiary items is available for use.

At the same time a somewhat smaller effort has been devoted to improvising and testing homemade or field expedient devices and techniques for accomplishing similar results.

Since the manufactured, precision devices almost always will be more effective, more reliable, and easier to use, why spend time on improvising field expedients to do the job?

For one thing shelf items will just not be available for certain operations for security or logistical reasons. In these cases the operator will have to rely on materials he can buy in a drug or paint store, find in a junk pile, or scrounge from military stocks.

Secondly, many of the ingredients and materials used in fabricating homemade items are so commonplace or innocuous they can be carried without arousing suspicion. The completed item itself often is more easily concealed or camouflaged.

In addition, the field expedient item can be tailored for the intended target, thereby providing an advantage over the standard item in flexibility and versatility.

While most of the pertinent information on sabotage shelf items is available in catalogues and other publications, much of the improvisation know-how has remained in the minds or files of a few individuals.

It is the intent of this manual to consolidate and bring up to date a selected body of this information and make it available for wider use.

The devices and techniques included have been selected because they have been well proved out and because they are practical, versatile, and not too difficult to do. Other techniques have been omitted because they are too unreliable or are just too hazardous to attempt except under the most carefully controlled conditions.

The techniques described are not so complex as to require a chemical laboratory or machine shop; however, many of them do assume access to basic demolitions and incendiary supplies, such as explosives, time (safety) fuse, detonating cord, blasting caps, and flammable materials, and to a few household tools.

This manual is not intended to stand completely on its own as a substitute for instruction, nor is it a text on basic demolitions. Its chief value should be to aid an instructor who already has a working knowledge of the subject in his training of action personnel. It also will serve as a refresher and guide to individuals who have been trained in these techniques but whose facility has declined from disuse.

EXPLOSIVES FOR SABOTAGE

Commercial and military explosives of one sort or another are in use around the globe. By far the most desirable ones for sabotage purposes are the military plastic explosives and TNT.

Plastic explosives vary in composition and go by different names. The U.S. military has two versions: C-3 and C-4. The British call theirs PE-2 and the Italians T-4. Other names used are RDX and, especially in Europe, Hexagen and Cyclonite.

TNT may be called by its full chemical name, Trinitrotoluene, or it may be referred to as Tolite, Trotyl, or TOL.

Both plastic explosives and TNT have the advantage of being very powerful while at the same time insensitive and quite safe to handle. Both also will retain their explosive properties after prolonged submersion and thus are excellent for underwater use.

Plastics have an additional characteristic which makes them ideal for many of the improvisations described later. They can be sliced or molded into any desired shape or pressed snugly against an irregular target.

TNT, on the other hand, usually is found in a rigid form -- either pressed or molded. Placement on target is therefore more difficult. Sometimes it is available in flake or powder form. Also, the block form can be safely pulverized. Some of the improvisations employ TNT in the loose state.

Dynamites also are used commercially worldwide and often may be more available than military explosives. They may be effectively employed in sabotage, although they are less potent, less versatile, more sensitive to shock or friction, and may not be waterproof.

If no other sources exist, artillery and mortar shells, bombs, land mines, and other munitions may be cannibalized for the high explosives they contain. These munitions usually contain cast explosives, which require extra strong blasting caps or boosters to detonate. However, some of the simpler cannibalization techniques will be discussed.

How about making your own explosives?

This can be done, and literature is available giving formulas and procedures. Unfortunately, the ones easiest and safest to make are the weaker ones and the least effective for most sabotage purposes.

One homemade explosive, using nitrate and oils, will be described.

STANDARD FIRING SYSTEMS

General

Since "good" priming often means the difference between a successful shot and a misfire, or between a low order and a high order detonation, a brief summary of the precautions to take and the pitfalls to avoid might be useful before proceeding to actual improvisations.

Explosives may be detonated electrically or nonelectrically. A nonelectric firing system will consist of a blasting cap, a length of time (safety) fuse or a firing device attached to the cap, and a means of activating the system: match, fuse lighter, delay mechanism, or trip wire. The electric system requires an electric cap which has two wires attached, perhaps additional wire, and a battery or batteries to provide current which activates the cap.

The nonelectric firing system is simple and requires a minimum of equipment. Once initiated, however, it passes from the control of the operator. The electric system can be retained in the control of the operator up to the actual firing. Electric caps also are more waterproof. On the other hand, the electric system is more complicated, employs more equipment, and may require the operator to remain near the scene at the time of firing. In addition, electric systems can be accidentally activated by static electricity and are hazardous to use in some target situations.

With either system, the location of the detonator in the charge has considerable effect on the performance of many devices or techniques. The business end of the blasting cap must be surrounded by the explosive to be detonated and, if plastic, the explosive should be firmly pressed around the cap.

Nonelectric Firing

Time fuse, or safety fuse, which is used in most nonelectric firing systems, is available in a multitude of varieties, both in color and rate of burning.

The only sure way to determine the burning time of any kind, or even roll, of safety fuse is to cut a measured length of it, light it, and time it. The operator can then select a length which will provide the time delay he desires.

Safety fuse is quite susceptible to moisture damage, so it is always a good idea to cut off and discard the first 6 inches or so from the exposed end of a roll.

By all odds, the most frequent cause of misfires in nonelectric systems is failure to seat the blasting cap firmly on the time fuse. Here's the proper way to do it:

After testing the burning rate of the safety fuse being used, cut a desired length of it with a sharp knife or with crimpers. Smooth and compress one end of the fuse by rolling the end between thumb and forefinger. This

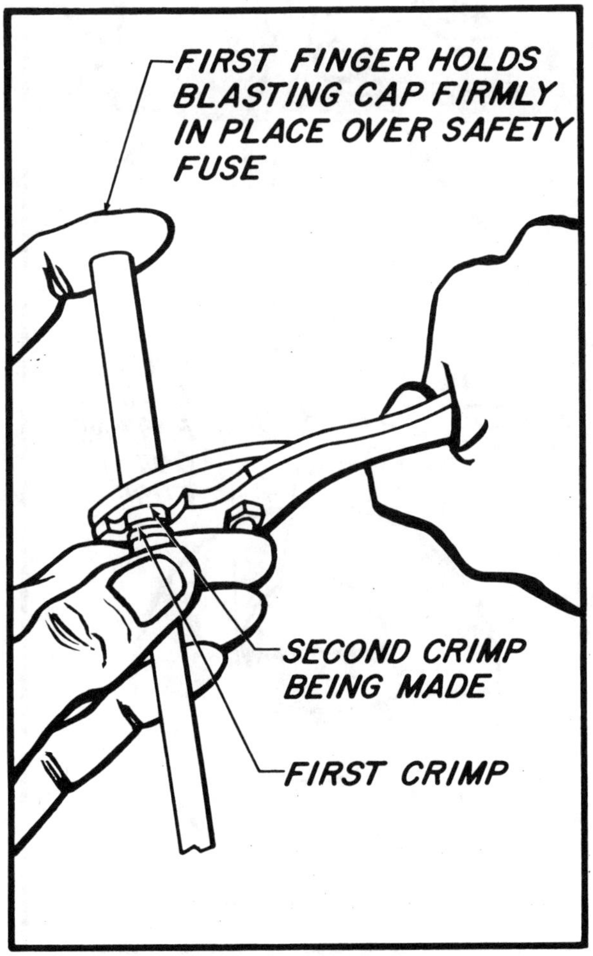

Fig. 1 — Crimping

will make it easier to insert into the cap. Next, insert the prepared end of the fuse firmly into the open end of the blasting cap until it stops. Hold the cap vertically, with the fuse pointing down, and keep it tightly seated by pressing down with the forefinger while gripping the fuse with other fingers of the same hand (see Figure 1). Now crimp the cap to the fuse at a point near the open end of the cap.

The easiest way to light safety fuse is with a mechanical fuse lighter, widely used in military services, or with special hot-burning fuse lighting matches. If ordinary matches must be used, here's the surest way to do it:

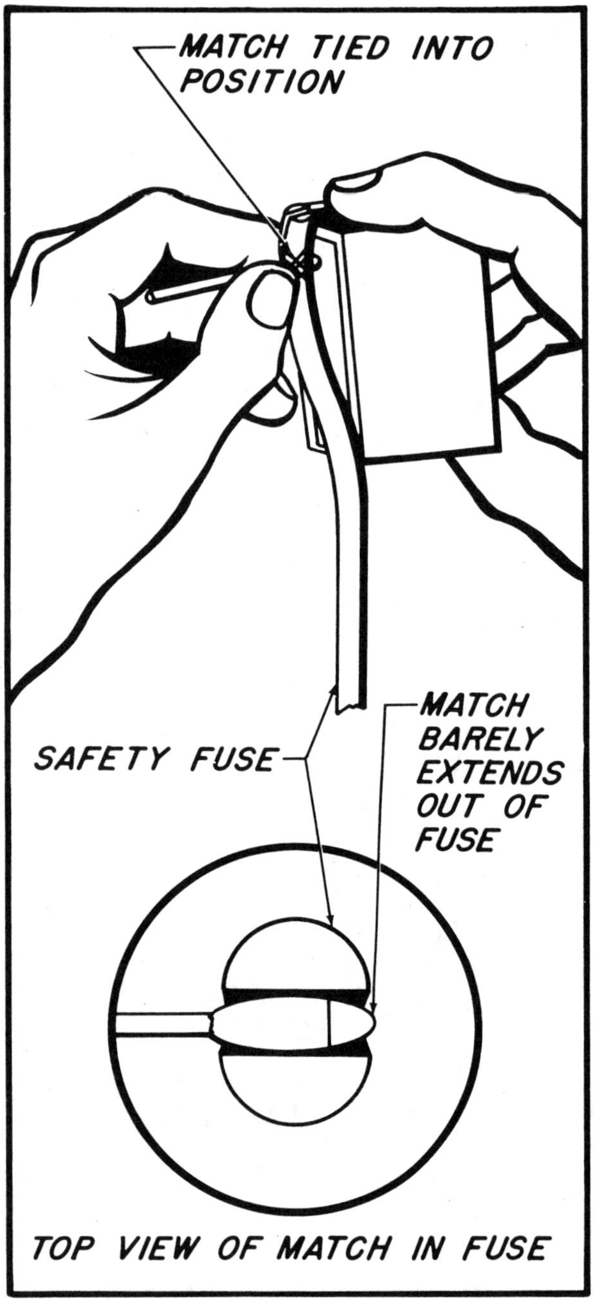

Fig. 2 — Lighting Safety Fuse

Make a half-inch split in the end of the fuse through the center of the powder train. Break off a match head and insert it into the split so that the match head is in intimate contact with the powder but with its tip protruding sideways from the split. The match head should be firmly held or tied in this position (see Figure 2) for igniting against the match-box or folder striker board.

Electric Firing

The most frequent troubles in electric firing are insufficient battery power, or shorts or breaks in the circuit.

An ordinary flashlight battery is strong enough to fire a single electric blasting cap if the current has to travel only through the cap wires or a short length of additional wire. The amount of extra power required will depend upon the type and length of firing wire, the number of caps and the method of connecting them.

Electric firing circuits always should be tested for breaks and sufficient power before the blasting cap is attached. This can be done by touching the bare ends of the two wires, which will be later connected to the blasting cap, to the bottom and metal side of a flash-light bulb. If the bulb lights up the circuit is good and the battery strong enough. If it glows only faintly or not at all the system probably will not work.

After testing, always be sure the power source is disconnected before attaching the blasting cap.

Primacord

Primacord, or detonating cord, differs from time fuse in that it detonates practically instantaneously along its entire length rather than burning progressively and relatively slowly.

Its chief use is to tie together separate charges for simultaneous detonation.

Since the shock of a single strand of primacord is not powerful enough to detonate most explosives, a knot must be tied at the end to

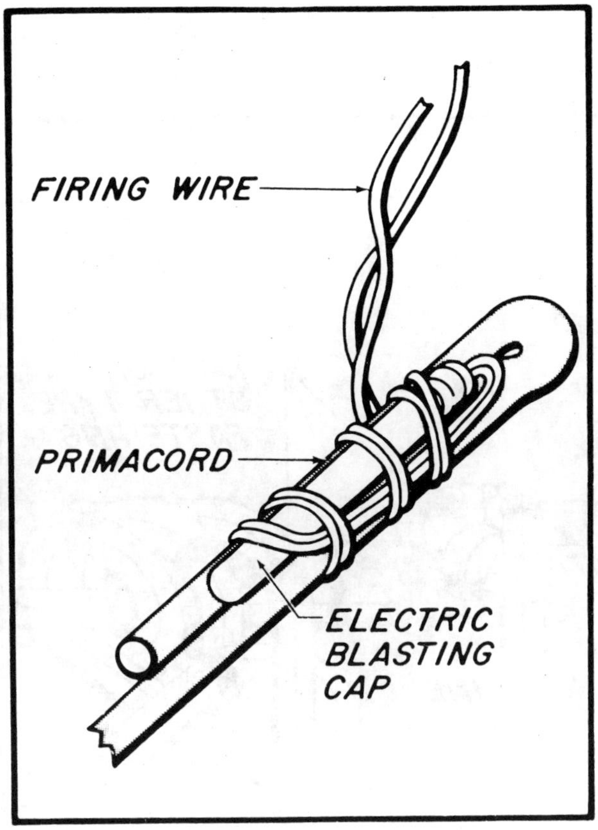

Fig. 3 — Electric Cap and Primacord

concentrate explosive force within the charge to be detonated. A properly tied knot, securely buried, is sufficient to detonate plastic explosives.

A blasting cap is necessary, however, at the point of initiation of the firing system to detonate the cord itself. The shock will then be carried to each charge attached to the system. The blasting cap always should be securely fastened, taped or tied, to the primacord, with the explosive end of the cap pointing down the line toward the charges to be fired (see Figure 3).

When laying out a firing system using primacord, avoid sharp angles or bends in the lines. The shock wave in the cord as it detonates will turn corners of 90° and continue. If the angle around a bend or to a branch line is sharper than 90°, the detonation may fail at that point.

A proper main line and branch line hookup is illustrated in Figure 4.

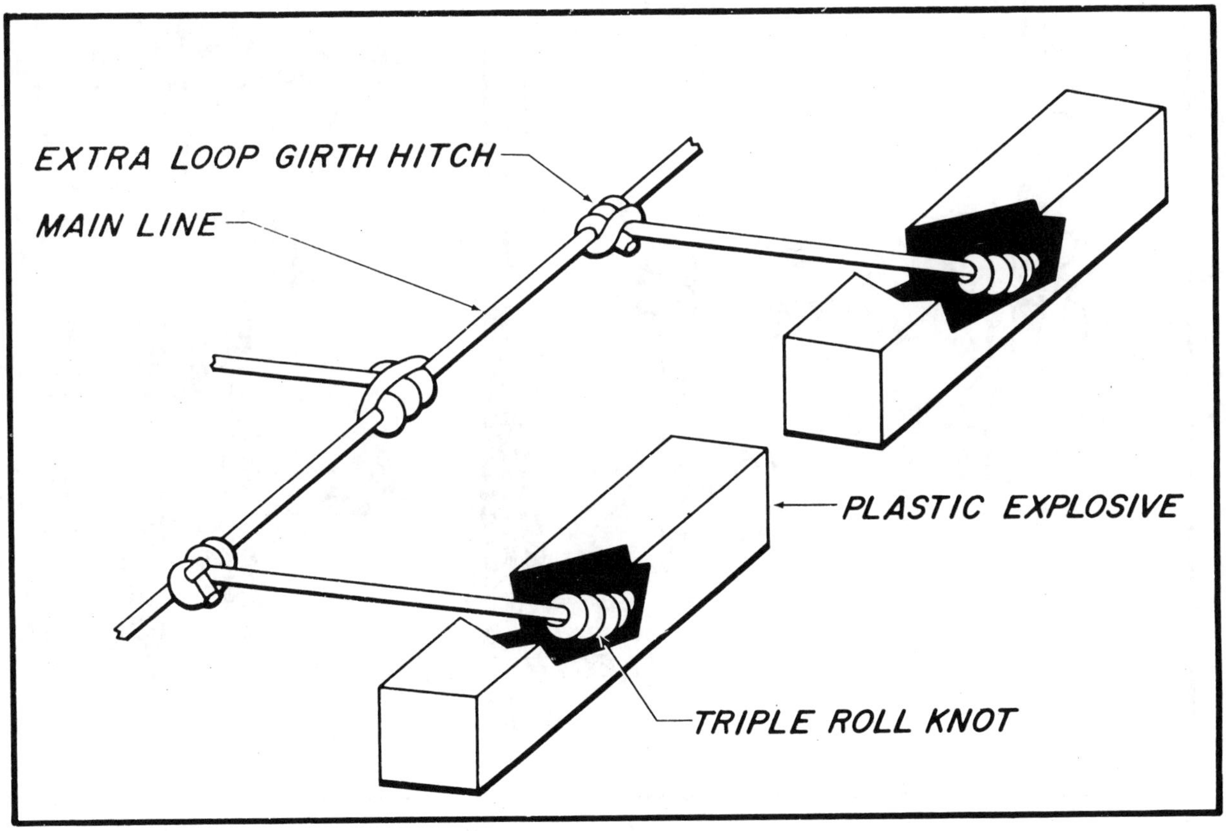

Fig. 4 — Primacord Branch Line System

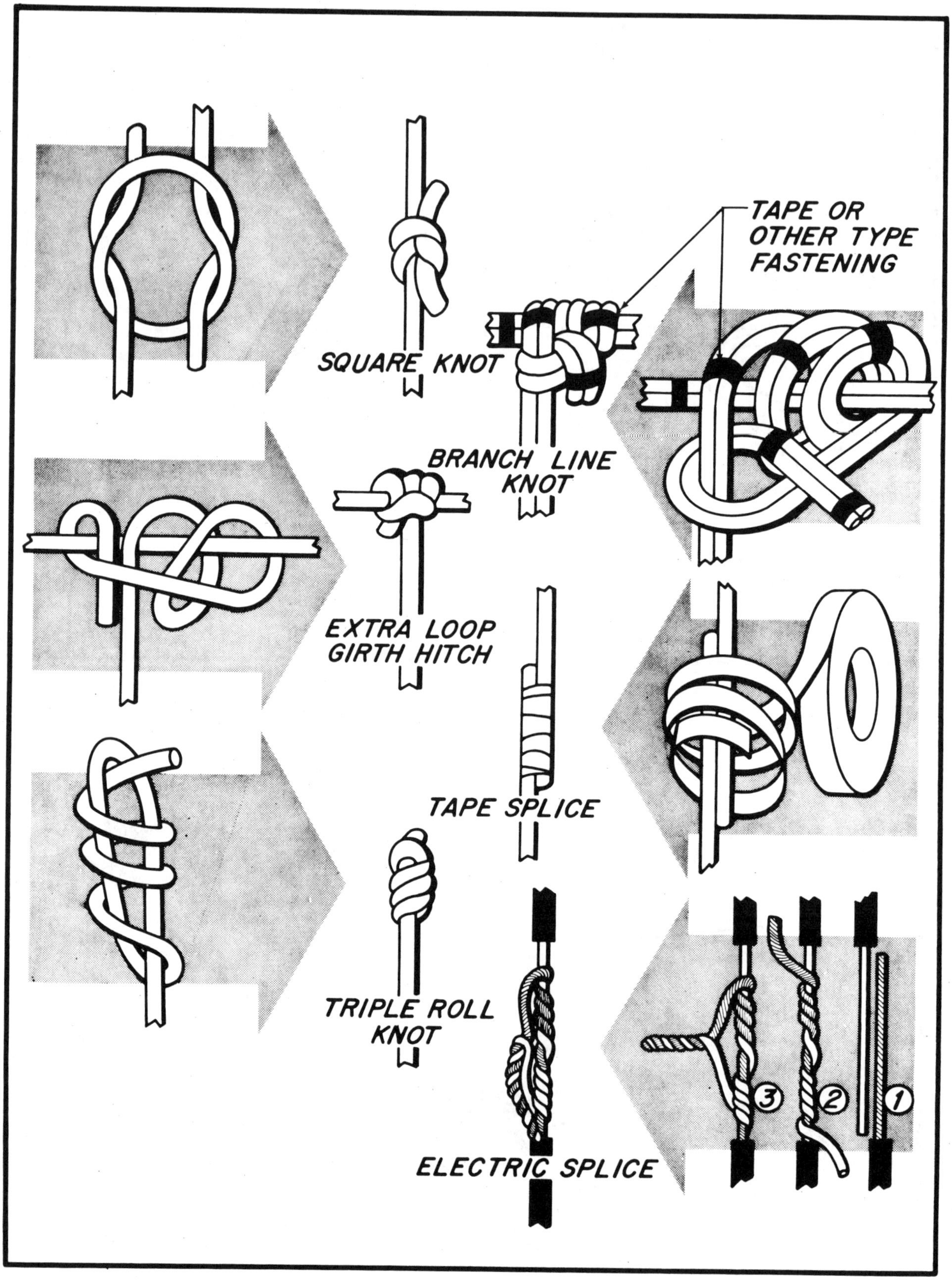

Fig. 5 — Splices and Knots

In priming plastic explosives with primacord, the triple roll knot (see Figure 5) is a good one to use. Three turns of the cord are concentrated in one tight knot, insuring detonation of the charge.

A good knot for tying a branch line to a main line is the girth hitch with an extra turn around the main line (see Figure 5). The knot should be pulled up snugly and the branch line extended at 90°.

Splices in primacord are best made with a square knot (see Figure 5). It may be a good precaution to tape or tie the knot to prevent loosening, particularly if the cord is stiff or slippery.

General

Depending upon the purpose and circumstances of the job to be done, the operator will need different ways of setting off his charges. He may need only a short delay, such as may be provided by a piece of time fuse, to permit him to get out of range of the explosion. He may need a reasonably accurate delay of several hours. He may wish to initiate the action instantaneously himself at a precise moment. Or he may wish to have the action initiated by the target.

A number of improvised electrical, mechanical, or chemical means of accomplishing these various purposes are described and illustrated in this section. The techniques presented illustrate different principles which can be employed and by no means exhaust the possibilities.

Electric Pressure Switch

This is a simple device to close an electric circuit and detonate a charge when pressure is applied by a target individual or vehicle.

MATERIALS: Two tin can lids or similar pieces of metal, dry paper, battery, electric wire, electric blasting cap.

PREPARATION:

1. Drive nail holes through one of two flat, shiny tin can lids so that sharp points of metal protrude from the other side.

2. Tightly attach a piece of hookup (small diameter electrical) wire to each lid through a hole punched near the edge.

3. Make a sandwich of the lids, with a piece of dry paper in between completely separating the lids. The lid with the nail holes may be on top or bottom, but the "pips" must point in toward the paper, as shown in Figure 6.

4. Test the device with a battery and flashlight bulb. Pressure on the top lid should push the sharp points through the paper to make contact with the lid below, thus completing a circuit. The amount of pressure

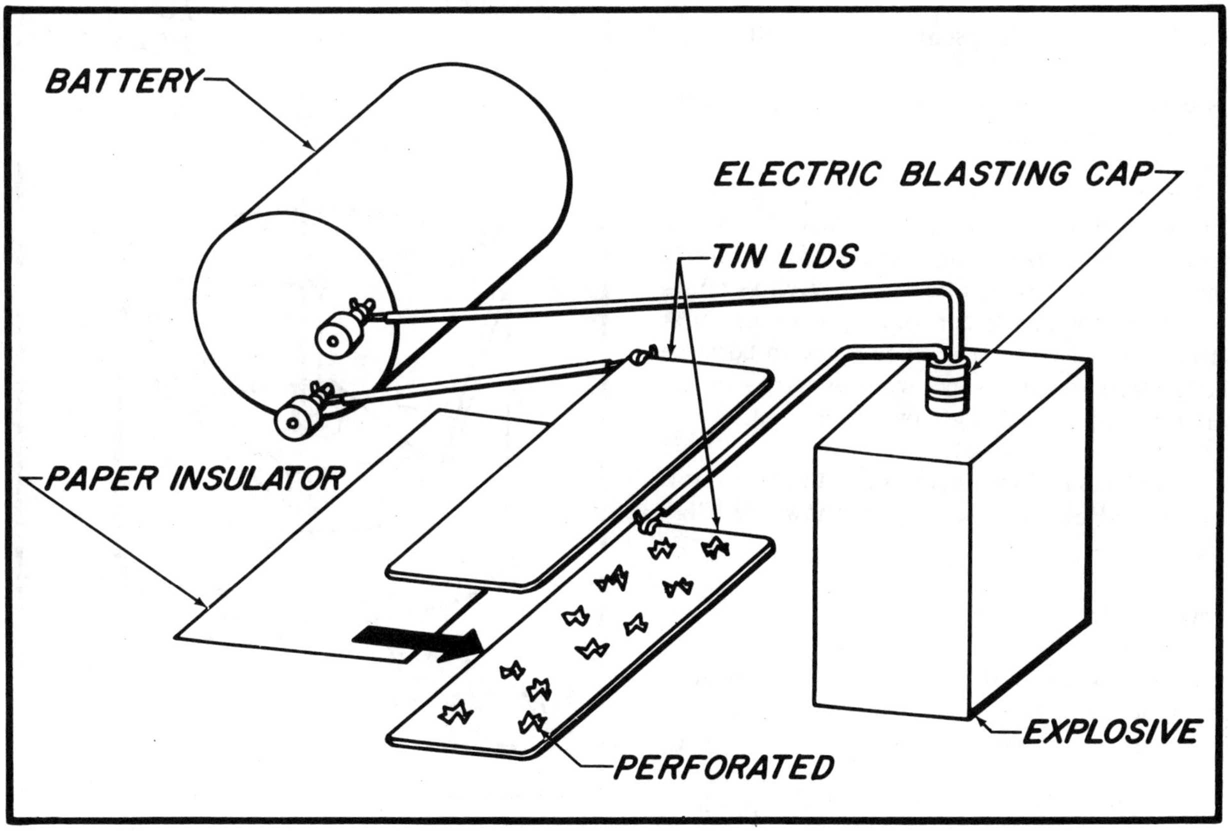

Fig. 6 — Electric Pressure Switch (Lids)

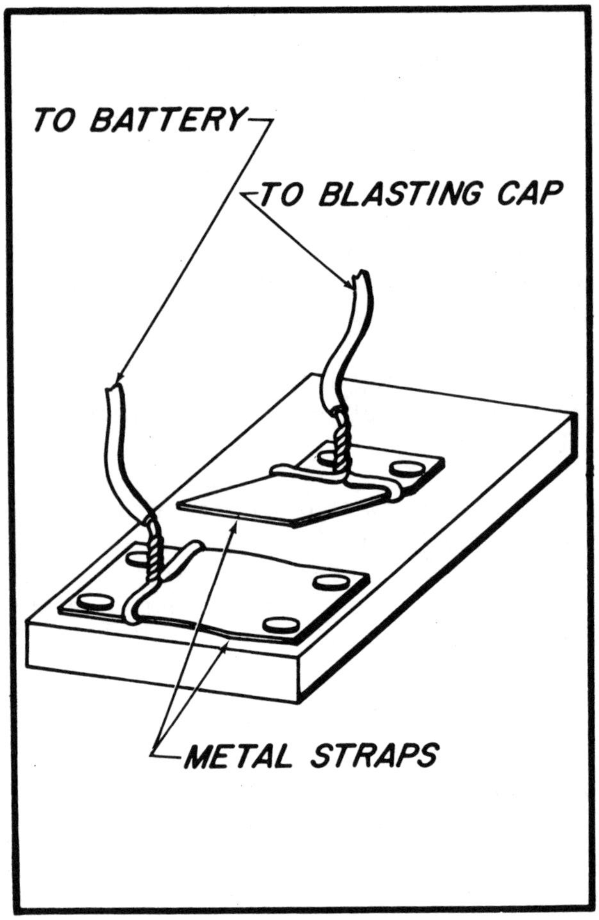

Fig. 7 — Electric Pressure Switch (Straps)

required can be varied by using paper of different types or thickness.

Before using the device in an operation, of course, the piece of test paper will have to be replaced by a fresh, unpunctured one of similar strength. Also, after the switch is positioned, and perhaps concealed, a further test with bulb and battery is necessary to be sure the switch is OPEN before the wires to the blasting cap and charge are connected.

A variation of this system, using two metal straps nailed to a board, is shown in Figure 7.

Water-Drip Electric Delay

This device will close a switch after a short delay. The delay time can be accurately controlled by the amount of water used.

MATERIALS: Tin can, wood scrap, electric wire, battery, electric blasting cap.

PREPARATION:

1. Make a very small hole in the side of the can at the bottom.

2. Use a small piece of wood for a float in the can. Attach a piece of stiff insulated hookup wire to this float so that the wire will extend 2 or 3 inches above the top of the can when the float is resting on the bottom (see Figure 8).

3. Fasten a second piece of insulated electric wire across the middle of the top of the can. Make a loop in the center of this wire and strip the insulation from the loop.

4. Now run the straight wire from the float through the loop of the wire crossing the top of the can. With the float just off the bottom of the can, mark the float wire at the point where it goes through the loop. Remove the insulation from this point on the wire to the top end.

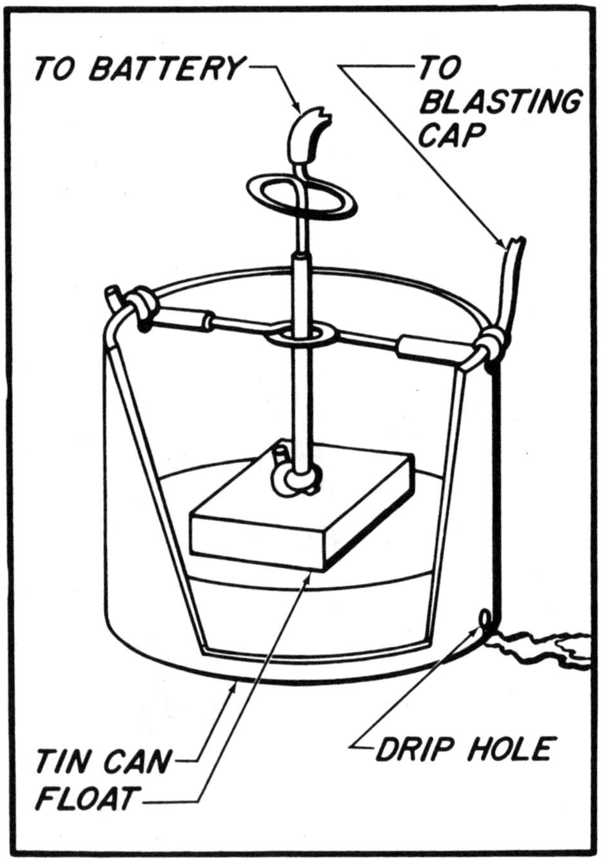

Fig. 8 — Water-Drip Electric Delay

5. Bend a kink or loop in the float wire at the point from which the insulation was removed. This bend should be made so that no matter how the float sinks to the bottom of the can, the bare wire at the bend will contact the uninsulated wire of the loop. Leave enough wire above the kink or loop to attach a hookup wire.

6. Check the operation of the device with a battery and flashlight bulb and determine the proper height to which water should be added in the can for the desired delay. A piece of tape or wax may be used to close the hole and prevent the water from dripping until it is desired to start the delay.

To use, place the delay on a firm, level support and fill with water to the desired level. Check to see that the circuit is open before making the final connection between the battery and blasting cap. After making the final connection, open the drip hole to start the delay.

Expansion Electric Delay

This is a variation of the drip delay and may be more suitable in a situation where leaking water is undesirable. The delay time is not readily varied but normally runs from 1/2 to 2 hours.

MATERIALS: Small bottle or test tube, dried peas, beans, or grain, small piece of cork or wood, battery, electric wire, electric blasting cap.

PREPARATION:

1. Cut a float from cork or wood to a slightly smaller diameter than the test tube or other container to be used.

2. Strip the end of a piece of insulated electrical wire and fasten the bared portion to the top of the float as shown in Figure 9.

3. Pour the required amount of peas, beans, or grain into the container and insert the float as illustrated.

4. Fasten a second hookup wire to the top of the container with a bare section extending across the center of the opening, as shown in Figure 9.

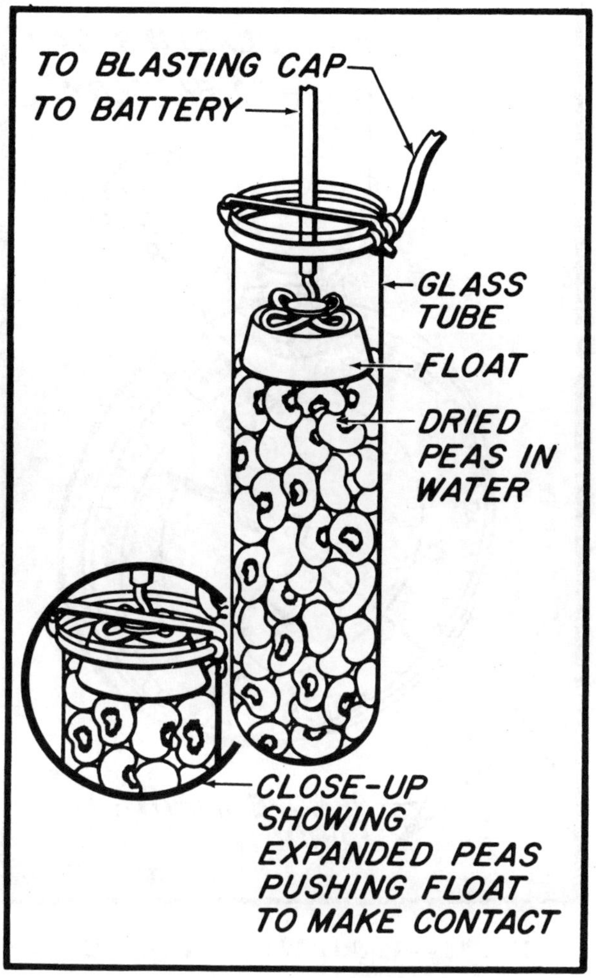

Fig. 9 — Expansion Electric Delay

5. Check the operation of the device by filling the container with sufficient water to cover the dry fill material. As the dried material swells it will push the float to the top of the container and complete the circuit.

6. Test the circuit with a flashlight bulb.

When used in an operation, the container should be securely fastened upright to a stable object. Fresh dry fill material will be required and, as always, the circuit should be tested to see that it is open before the blasting cap is attached.

Pocket Watch Electric Delay

Provides an accurate time delay of up to 11 hours.

MATERIALS: Pocket watch with plastic crystal, small screw, electric wire, battery, electric blasting cap.

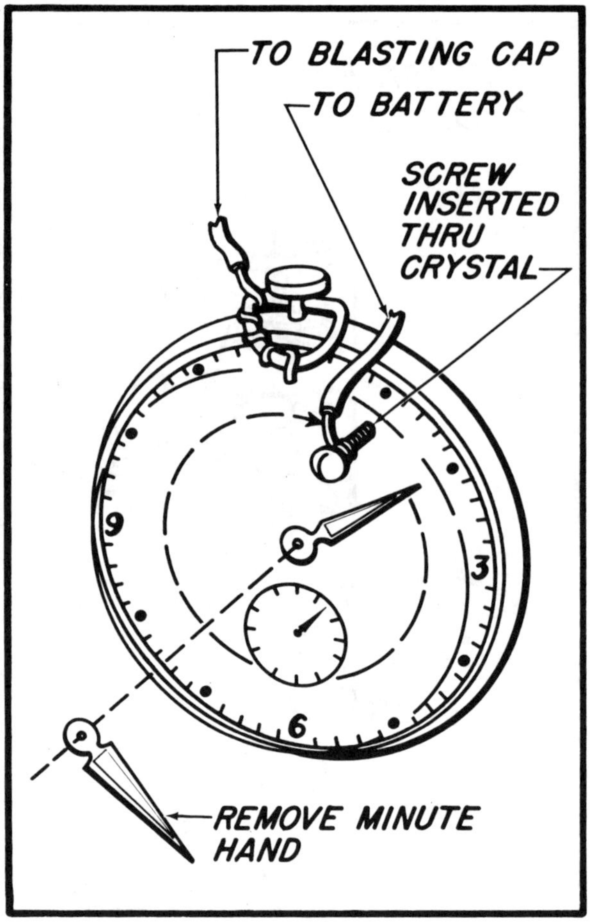

Fig. 10 — Pocket Watch Electric Delay

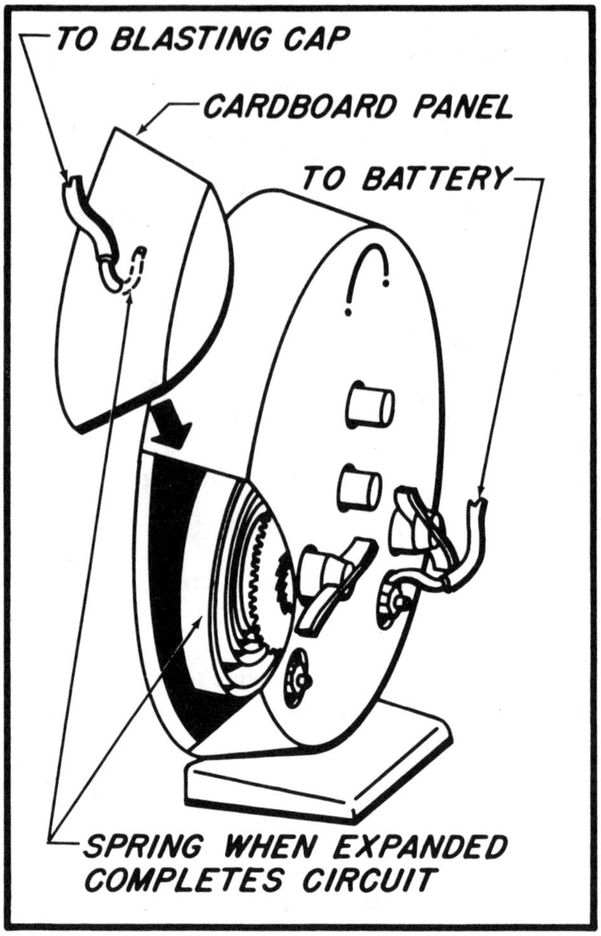

Fig. 11 — Seven Day Electric Delay

PREPARATION:

1. Remove the crystal from the watch.

2. If a delay of less than 55 minutes is desired, remove the hour hand; otherwise, remove the minute hand. (The hand may be broken off, if care is taken not to dislodge the other hand.)

3. Make a small hole in the crystal slightly less than the length of the hand from the center of the crystal. Insert the screw from the front of the crystal and tighten it so that it projects from the concave side. Carefully scrape any paint from the edge of the hand where it will make contact with the screw. Replace the crystal in the watch. See Figure 10.

4. Now turn the hand to see that it contacts the screw. If the hand passes beneath the screw, tighten the screw until it obstructs passage of the hand. However, be sure the screw does not contact the face of the watch (which may be metal).

5. Attach one length of hookup wire to the screw head and another to the winding stem or any other bare metal part of the watch case.

6. Test the device for proper functioning and accuracy of delay.

To use, wind the watch fully and be sure it is running. Set the desired time delay by backing the hand away from the contact screw. Check to be sure the circuit is open before making final connections to battery and blasting cap.

Seven Day Electric Delay

Provides an accurate delay of up to seven days.

MATERIALS: Seven day clock, cardboard or other insulating material, tape or glue, electric wire, battery, electric blasting cap.

PREPARATION:

1. Remove the back of the clock and cut away that part of the housing next to the clock mainspring. See Figure 11.

2. Replace the clock housing and attach one hookup wire tightly to any bare metal part on the outside of the clock.

3. Make a panel from cardboard or other stiff nonconducting material to close the opening in the housing. This should be firmly placed so as to obstruct further expansion of the clockspring as it unwinds.

4. Pierce the cardboard with the bare end of a hookup wire so that the clockspring will make contact with the end of the wire when it unwinds.

5. To preset a desired delay, first wind the clock fully. Push the hookup wire in through the cardboard until the bare end touches the clock spring. As the spring expands, it will push the wire back out through the opening. When the desired delay in days and hours has elapsed, firmly secure the wire at that point with glue or tape. To prevent the preset delay from being dislodged by further pressure from the clockspring, wind the clock a few turns and then stop the clock's movements until it is needed.

To use, wind the clock fully and be sure it is running. The position of the hands does not matter since the delay is based on the unwinding of the spring. Check to make sure the circuit is open before making final connections to battery and blasting cap.

Electric Pull-Release

Initiates an explosion when a trip wire is bumped into, cut or broken.

MATERIALS: Hacksaw blade, table knife or strip of springy steel, board, nails, electric wire, battery, electric blasting cap.

PREPARATION:

1. Anchor the blade or knife firmly at one end of a board, so that it is resting on edge and extending horizontally down the board.

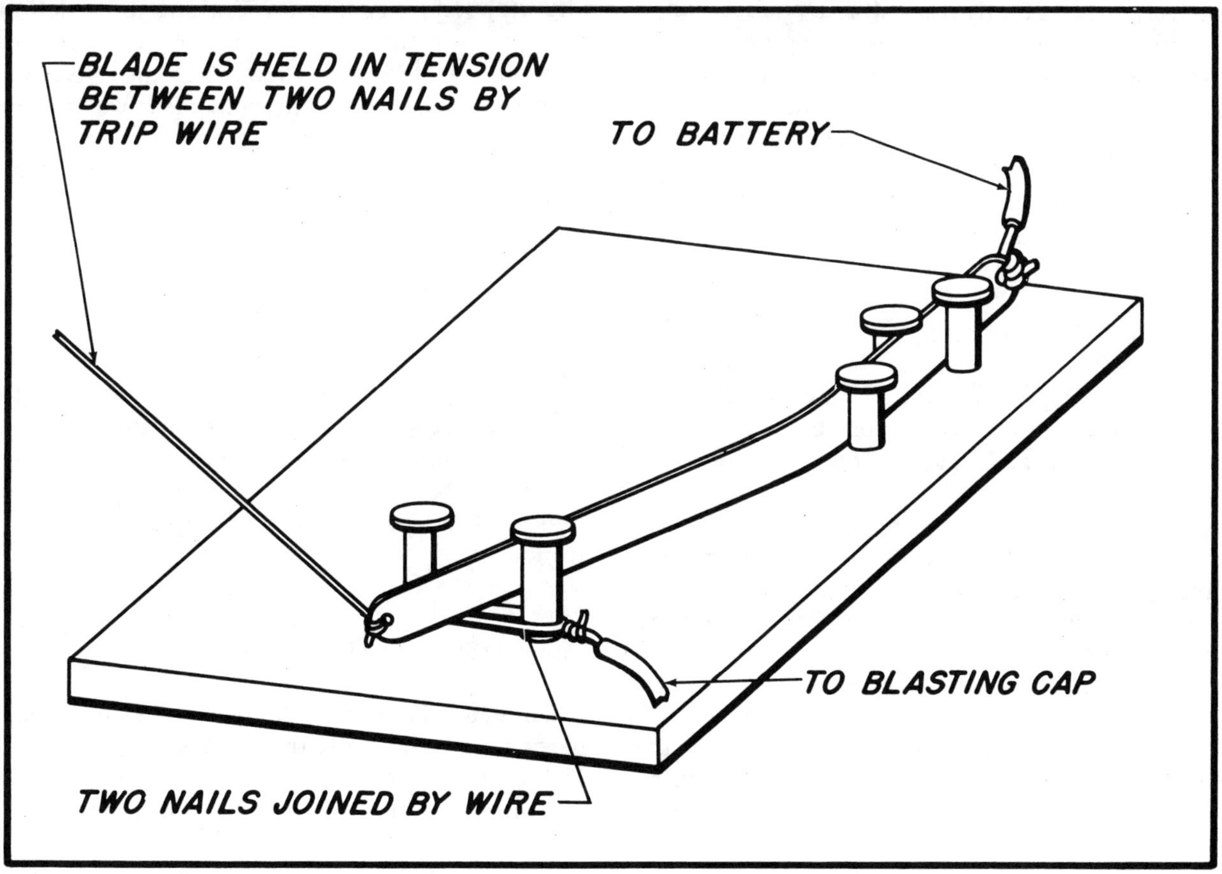

Fig. 12 — Electric Pull-Release

2. Flex the tip of the blade in either direction and drive a nail at that point outside the blade but touching it. Drive a second nail an inch or so (depending upon the flexibility of the blade) on the other side of the blade. If the blade is now released it should rest against the second nail in a slightly flexed condition. See Figure 12.

3. Connect the two nails with a short piece of uninsulated wire so they become a common contact. The connecting wire must be as low on the nails as possible so the blade can swing between the nails and above the wire without touching it.

4. Attach one length of hookup wire to the fixed end of the blade and a second piece to either of the nails.

5. Now test the circuit with battery and flashlight bulb. If the blade is held or secured midway between the two nails the circuit should be open. If the blade is moved to touch either of the nails, the circuit should close.

To set up the device, the mounting board should be firmly fixed or anchored. Attach the trip wire firmly to a fixed position and fasten the other end to the movable end of the hacksaw blade or knife. Adjust the tension of the trip wire until the blade is held midway between the two nails. Check to be sure the circuit is open before making final connection to blasting cap. If the trip wire is pulled or broken the circuit will be closed, firing the blasting cap.

Electric Tilting Mercury Switch

A booby trap device which may be activated by a pull wire or by tilting of object to which applied, such as luggage, door handle, or hinged cover of box.

MATERIALS: Mercury, eye dropper or small glass tube, cork or wood, cement or glue, electric wire, battery, electric blasting cap.

PREPARATION:

1. Seal the narrow end of an eye dropper in an open flame and remove the squeeze bulb from the other end.

2. Pour a few drops of mercury into the dropper.

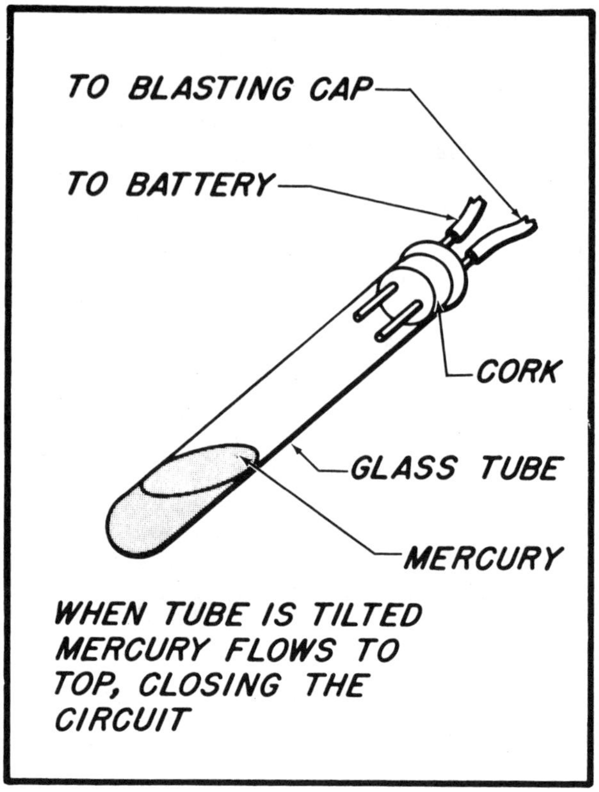

Fig. 13 — Electric Tilting Mercury Switch

3. Fit a cork or wood stopper tightly into the dropper. Make two small holes through the stopper or two grooves along the sides of the stopper through which the bare ends of two pieces of hookup wire can be inserted (see Figure 13). When the mercury is tilted toward the wires it will complete an electrical circuit. When it is tilted toward the other end the circuit will be open. A trip wire also may be used to activate the device.

4. Test the circuit and operation of the device.

To use, secure the device to the object to be booby trapped, with the mercury drained away from the ends of the hookup wires. Check to be sure the circuit is open before making final connections to battery and blasting cap.

Mousetrap Electric Release

This is a booby trap firing device which may be concealed in a cigar box, book, package or under any object heavy enough to hold the arm of the mousetrap in a cocked position.

MATERIALS: Spring mousetrap, electric wire, battery, electric blasting cap.

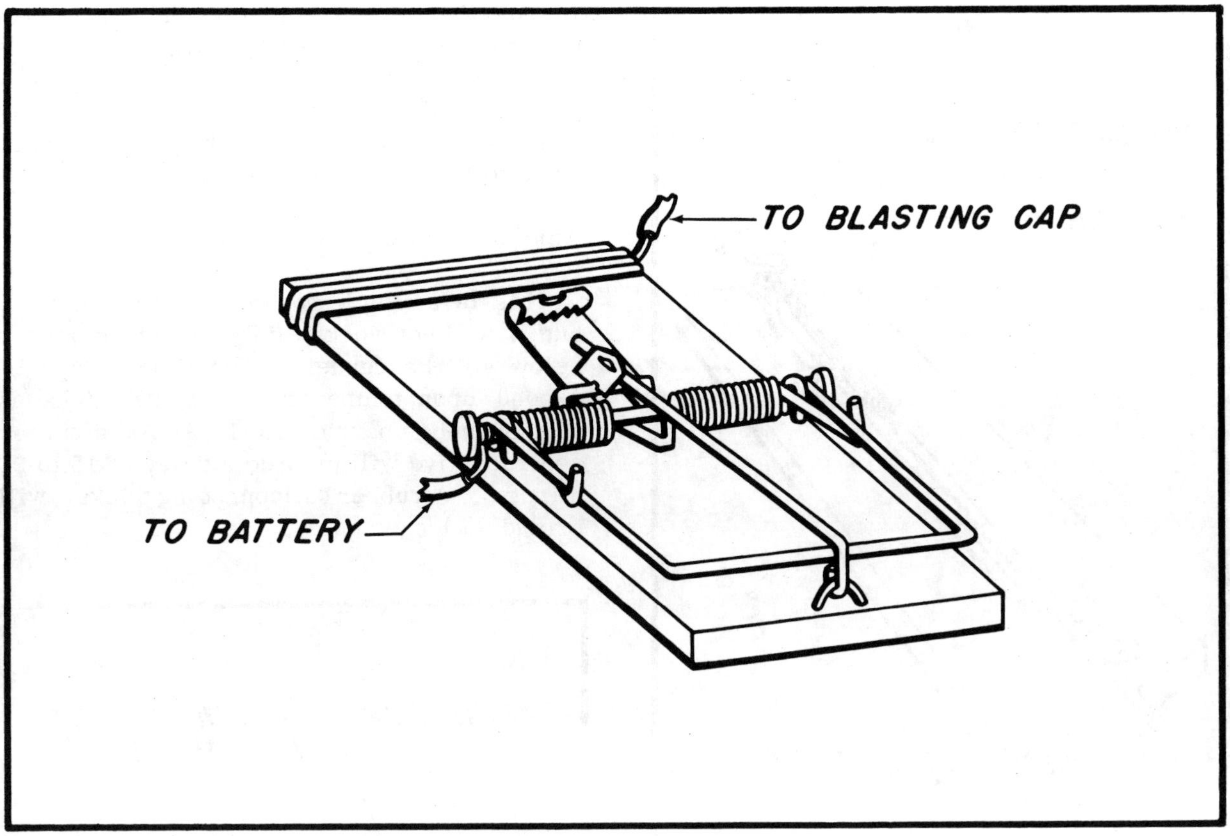

Fig. 14 — Mousetrap Electric Release

PREPARATION:

1. Attach one length of hookup wire firmly to the spring which drives the "catcher" arm of the mousetrap.

2. Strip the insulation from about 6 inches of a second hookup wire and wrap this bared portion tightly around the wooden end of the mousetrap where it will be contacted by the catcher arm when released (see Figure 14).

3. Check the circuit and the operation of the device.

To use, fasten the device securely in place and cock the trap, making sure the lid or weight being used will hold the arm down fully in the cocked position. Check to be sure the circuit is open before making final connections to battery and blasting cap. Removal of the lid or weight will spring the trap and fire the blasting cap.

Electric Pull-Slide Switch

This simple item provides perhaps the best and easiest method of firing a charge electrically by means of a pull wire. It may be rigged so the target trips the wire or it may be controlled by the operator from a safe distance. It is an excellent device to use with bazooka rockets against trucks, tanks, or locomotives.

MATERIALS: Rubber band or other piece of elastic, stick, trip wire or string, electric wire, battery, electric blasting cap.

PREPARATION:

1. Cut two 6 inch lengths of hookup wire and make a 1/4 inch wide loop of bare wire in one end of each.

2. Bend the loops at right angles to the wire and then thread the two wires together by passing the straight end of each wire through the loop in the other. The bare loops should now make contact if the wires are pulled in opposite directions. See Figure 15. (Although the device can be used in this simple condition, it will be more reliable if the following additional steps are taken.)

3. Attach a rubber band to one of the loops and fasten the other end of the rubber band

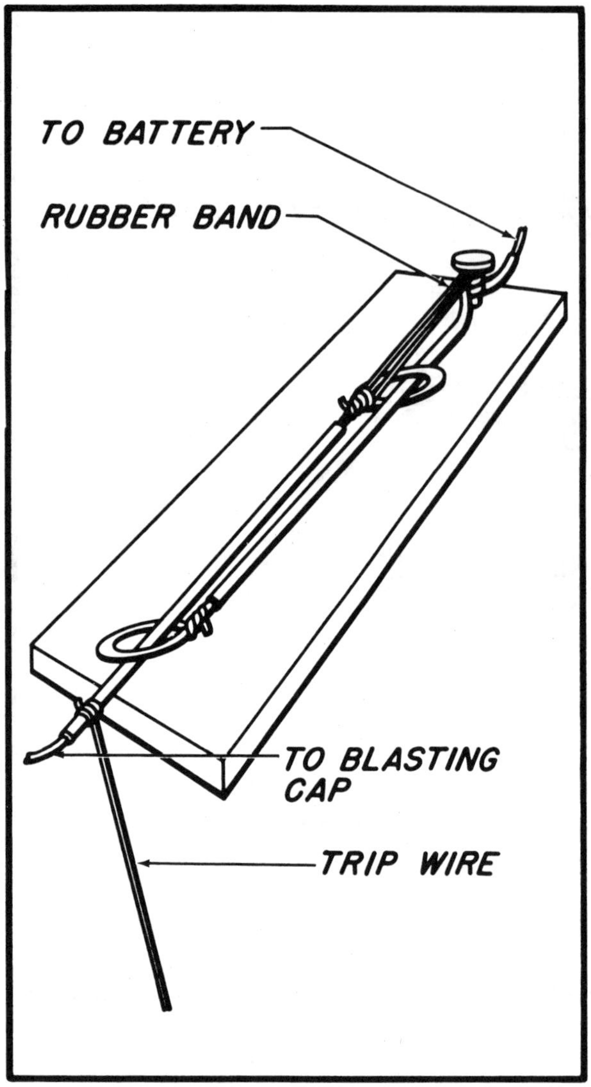

Fig. 15 — Electric Pull-Slide Switch

to the end of a stick, which can be about 10 or 12 inches long. To the same end of the stick, tie down the end of the wire which passes through the loop to which the rubber band is attached. An inch or more of this wire should extend beyond the end of the stick and the two loops should now be 2 or 3 inches apart. See Figure 15.

4. Attach the trip wire or string near the straight end of the hookup wire to which the rubber band is fastened. A pull on the trip wire should now stretch the rubber band and bring the two bare loops into contact.

5. Test the circuit and operation of the device.

To use the device, fasten the stick securely to a fixed object near the charge to be fired. Locate the trip wire in such a manner that the intended target will trip over it, or run it to a concealed position where it can be controlled by the operator. Check the circuit to be sure it is open before making final connections to battery and blasting cap.

Acid and Rubber Chemical Delay

This is a simple nonelectric delay device, but it will not work reliably at temperatures below 40° Fahrenheit. The time delay will depend upon temperature and the thickness of the rubber barrier used. At 70° a rubber contraceptive will provide a delay of 15 to 20 minutes. A rubber balloon, being thicker, will provide a longer delay.

Fig. 16 — Acid and Rubber Chemical Delay

MATERIALS: Small container such as a paper cup or tin can, rubber contraceptive or toy balloon, sugar-chlorate mixture (see section on incendiaries), CONCENTRATED sulfuric or nitric acid, nonelectric blasting cap.

PREPARATION:

1. Put two spoonsful of sugar-chlorate mix into the container.

2. Carefully check the rubber membrane to be used to be sure it has no tiny pinholes through which the acid could leak. Fasten the membrane over the top of the container completely and securely, but leaving a pocket in the membrane deep enough to hold a spoonful or so of liquid. See Figure 16.

To activate the delay, first fasten the container securely in an upright position. Make a hole in the side of the container near the bottom and insert the open end of a nonelectric blasting cap. Be sure that some of the sugar-chlorate mix gets into the open end of the cap. The business end of the blasting cap can be inserted directly into the explosive charge or connected to the charge with primacord. To start the delay, carefully place a spoonful of concentrated acid in the pocket of the membrane -- preferably using an eyedropper. When the acid eats through the membrane, the resulting flame will be sufficient to fire the blasting cap.

If necessary, a paper barrier may be used in place of rubber, but in any case the material should be tested before using.

Chemical Instantaneous Initiator

A simple nonelectrical initiating device which can be activated by pressure or by a trip wire.

MATERIALS: Eyedropper or other small glass tube or vial, rubber contraceptive or toy balloon, sealing wax, sugar-chlorate mixture (see section on incendiaries), CONCENTRATED sulfuric or nitric acid, nonelectric blasting cap.

PREPARATION:

1. Seal the narrow end of an eyedropper in an open flame or with sealing wax and remove the bulb from the other end.

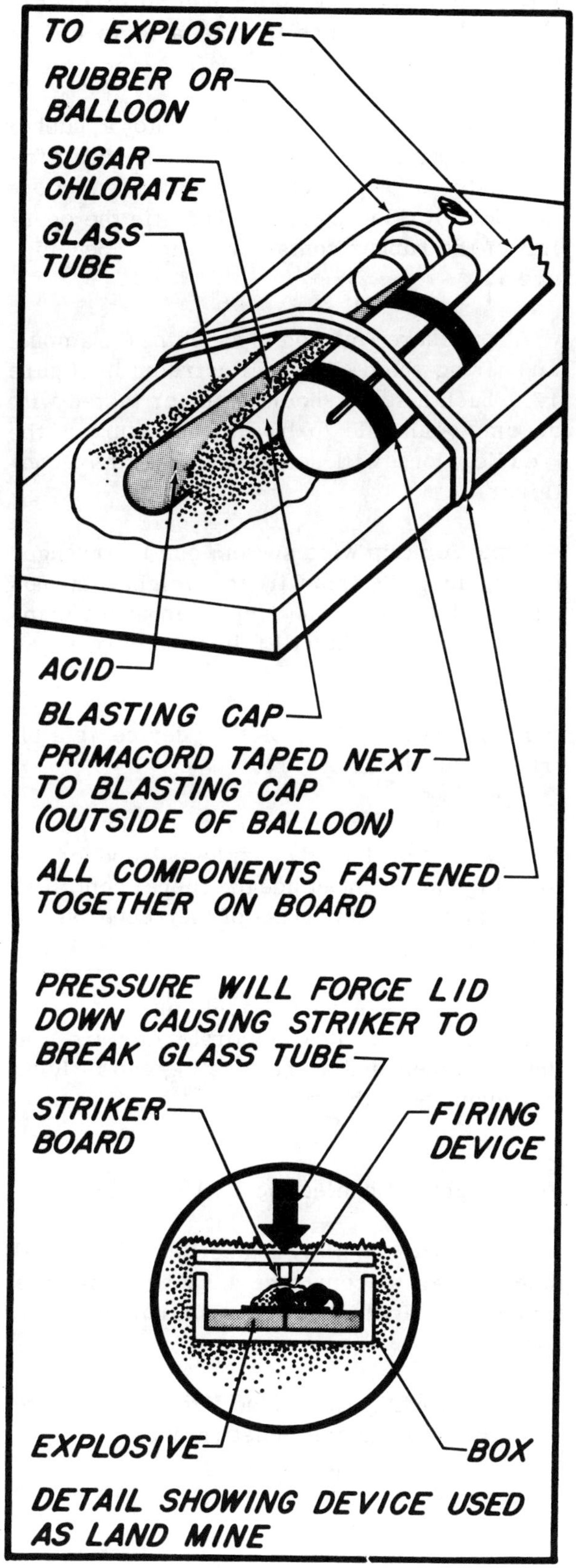

Fig. 17 — Chemical Instantaneous Initiator

2. Carefully fill the dropper at least half full of concentrated acid and seal the open end with wax. Let cool, then rinse and dry thoroughly to insure that no acid is on the outer surface.

3. Place the acid ampule into a rubber contraceptive or toy balloon, along with a few spoonsful of sugar-chlorate mixture and one nonelectric blasting cap. Then tie the opening of the rubber container tightly. See Figure 17.

A pressure method of crushing the ampule and firing the device is illustrated in Figure 17. Such a setup should be prior tested with an empty ampule to be sure pressure of the treadle alone will not crush the glass container.

A pull or trip wire system can be arranged by attaching the trip wire to a weight suspended or lodged above the ampule in such a manner that the weight will fall if the wire is cut or pulled.

Final positioning of this device is a bit tricky and requires care.

First, a length of primacord must be securely attached to the explosive end of the blasting cap from outside the rubber container so that only the thickness of the rubber separates the two.

Secondly, be sure the acid ampule is surrounded by the sugar-chlorate mixture and that the open end of the blasting cap is lying in a quantity of the mixture.

Fuse Lighter Nonelectric Pull

This device employs a modified mechanical fuse lighter to construct a booby trap or a firing device which can be controlled by the operator.

MATERIALS: M-2 fuse lighter, trip wire or string, nonelectric blasting cap.

PREPARATION:

1. Remove and discard the pull ring of the fuse lighter and ease the cotter pin partially

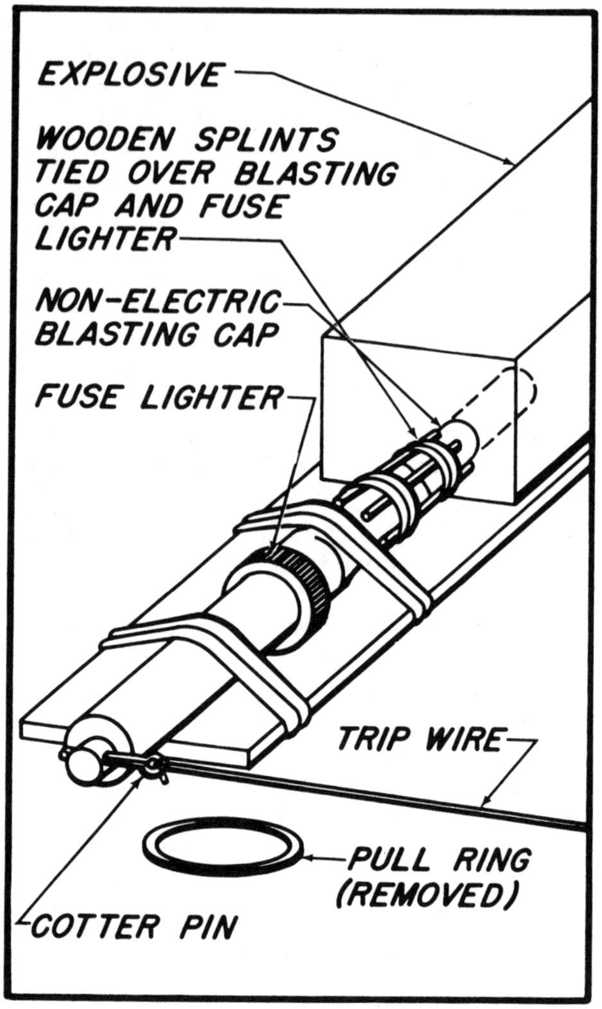

Fig. 18 — Fuse Lighter Nonelectric Pull

out until only the last 1/8 inch or so of the tips protrude.

2. Taper the open end of the blasting cap slightly with crimper or pliers and insert it firmly into the open end of the fuse lighter.

3. Fasten the cap securely in place with wood splints and string or tape as illustrated in Figure 18.

To use this device, the blasting cap may be inserted directly into the explosive or it may be connected to the explosive with primacord. In either case the firing device should be securely attached to a solid object. Run the trip wire from a fixed location to the cotter pin, leaving some slack in the line. A pull on the trip wire will remove the cotter pin, initiating the charge.

PRECAUTION: Since there is no safety pin in this device, beyond the cotter pin, extra care must be used to prevent accidental operation.

Fuse Lighter Nonelectric Pressure

This device employs a modified fuse lighter to initiate a booby trap when the target applies pressure -- such as in an antivehicle mine or a derailment system.

MATERIALS: Scrap wood and nails, M-2 fuse lighter, nonelectric blasting cap.

PREPARATION:

1. Follow steps 1, 2, and 3 and same precautions as in preceding technique.

2. Fashion a pressure board which, when moved downward, will pull the cotter pin from the fuse lighter. See Figure 19.

The blasting cap may be inserted directly into the explosive or a primacord connection can be used.

Fig. 19 — Fuse Lighter Nonelectric Pressure

IMPROVISED DETONATORS

General

One of the more difficult things to improvise in the field is a substitute for a blasting cap. This section will discuss two feasible techniques: one, the fabrication of a detonator by using a rifle, and two, cannibalizing the detonator from a fragmentation grenade.

Improvised Rifle Detonator

This technique provides a method of initiating primacord -- and thus any explosive charge -- if a regular blasting cap is not available. The device may be rigged as a booby trap with trip wire or it can be fired remotely by the operator. Since the technique does require the destruction of a usable rifle it likely would be used only when there is no other alternative. In addition, care should be taken in all phases of the preparation, because when the weapon is fired it will break into many fragments which would severely injure anyone holding the weapon or close to it.

MATERIALS: Expendable rifle, cartridge to fit rifle, primacord, trip wire or pull cord, wire or twine for lashing.

PREPARATION:

1. Check the rifle to be sure the safety and firing mechanism operate normally.

2. Remove the bullet from a live cartridge by gently loosening and pulling with pliers. Empty the powder from the cartridge and dispose of it by burning or save it in a moisture tight container for other uses.

3. Slit open about 1 yard of primacord with a sharp knife and collect the powder on a piece of clean paper or in some clean container. (NOTE: The explosive used in primacord usually is PETN. Other explosives sometimes used in detonating cord are Tetryl, RDX, and TNT. In this technique Tetryl will work as well as PETN. The others may not.)

4. Fill the empty cartridge case with the powder obtained from the primacord, packing it tightly with a pencil or stick. Do not pound or hammer on the stick. Leave the neck of the cartridge case empty.

5. Cut a length of primacord 1 yard longer than the barrel of the rifle. Insert one end into the empty neck of the cartridge. Hold the primacord firmly against the powder in the cartridge and carefully crimp the cartridge case neck around the primacord with pliers. Care must be taken in this step to insure that the primacord is held securely in the cartridge case, but that the neck of the cartridge is not so badly distorted by the pliers that it will not slip easily into the chamber of the weapon.

6. Check the fit of the cartridge by feeding the loose end of primacord down the weapon's barrel from the breech end and out the muzzle. By gently pulling the primacord through the barrel and guiding the cartridge into the chamber, determine whether the cartridge fits easily and whether the action of the rifle will close fully over it. Do not force the cartridge or it will be difficult to remove after this check. If it does not fit properly, remove it and adjust the crimp with pliers until the proper fit is obtained.

To use this device, firmly lash the unloaded rifle to a fixed object such as a stake or tree. See Figure 20. Tie a long pull cord or trip wire to the trigger and test the empty weapon once or twice to see that the trigger moves freely when pulled and that the firing pin snaps forward.

When the installation has been completed and checked, then load the primacord and detonator into the rifle as before -- leaving the action of the rifle open.

Connect the charge to be fired by tying the primacord leading from the charge to the end of the primacord extending from the muzzle of the rifle.

Before closing the action of the rifle, position the trip wire or pull cord, making sure there is enough slack to avoid tension on the trigger.

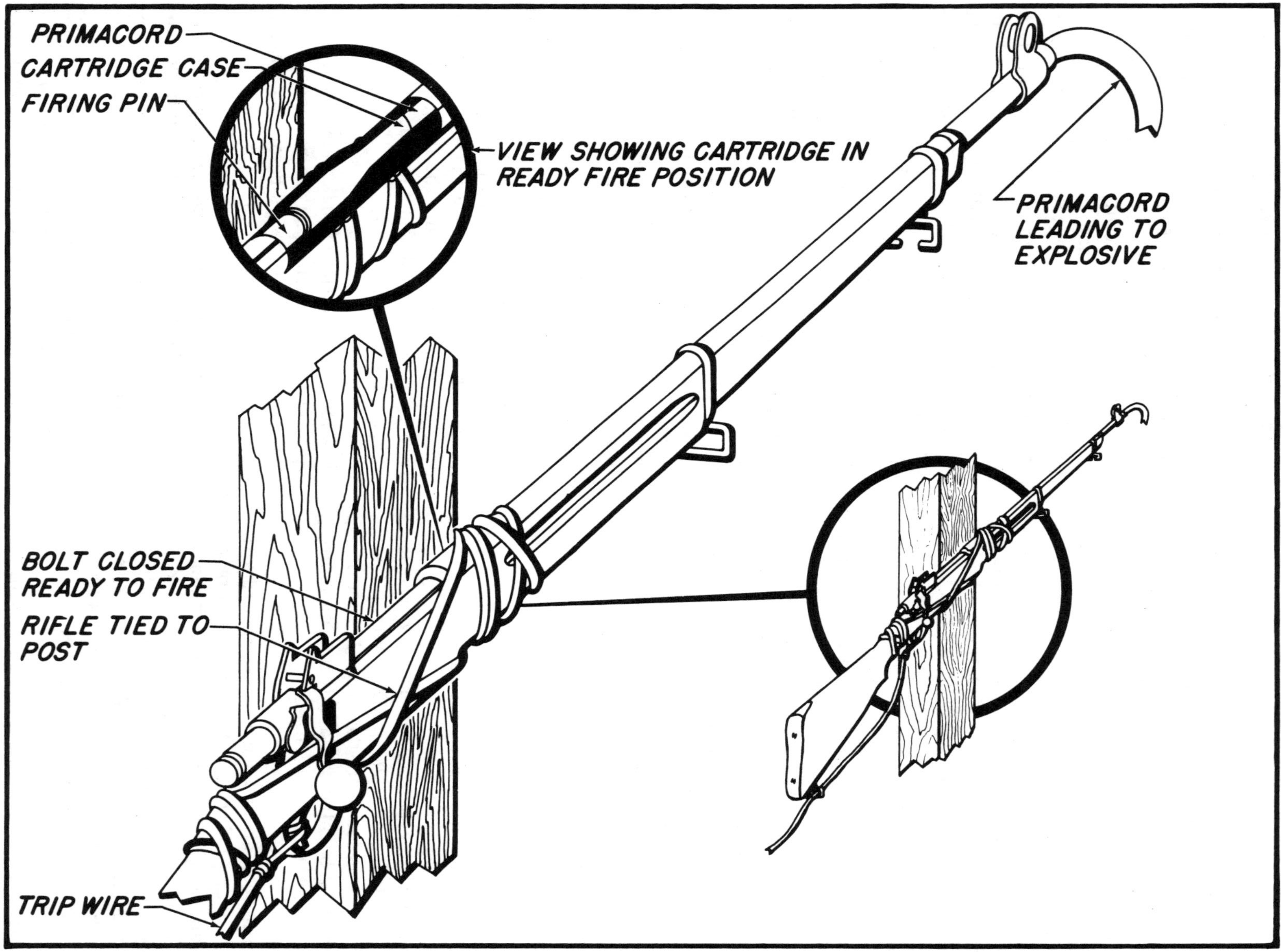

Fig. 20 — Improvised Rifle Detonator

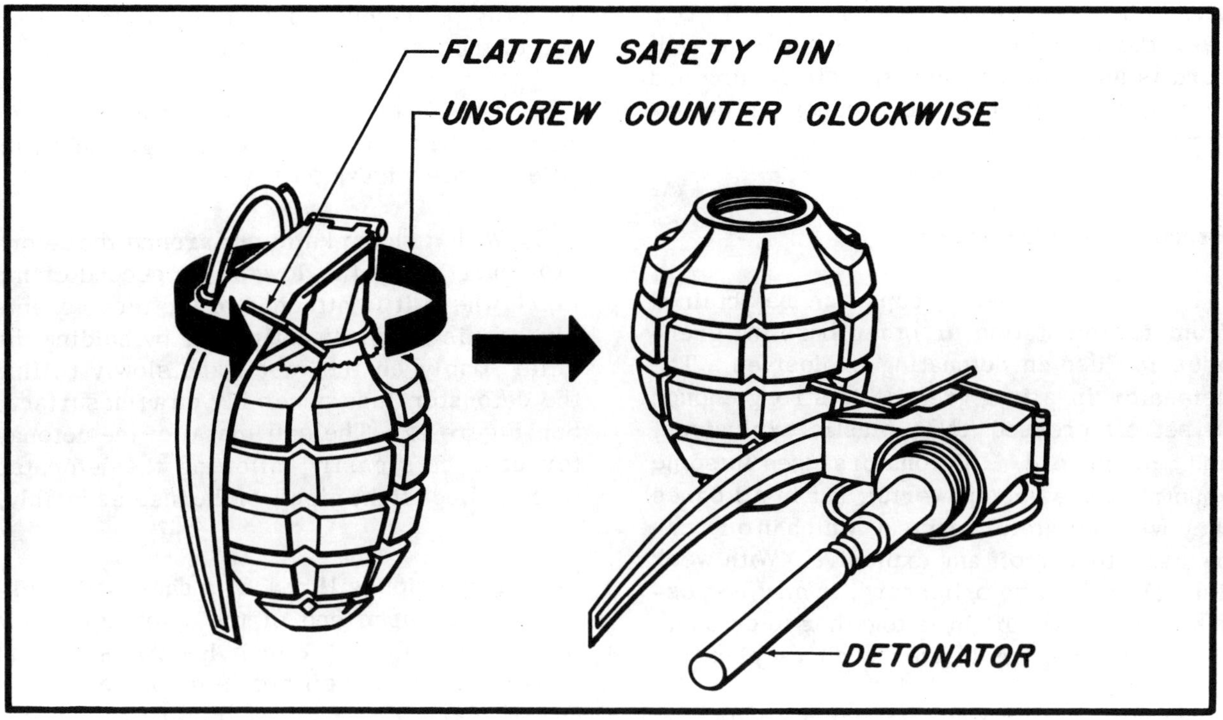

Fig. 21 — Removing Grenade Fuse

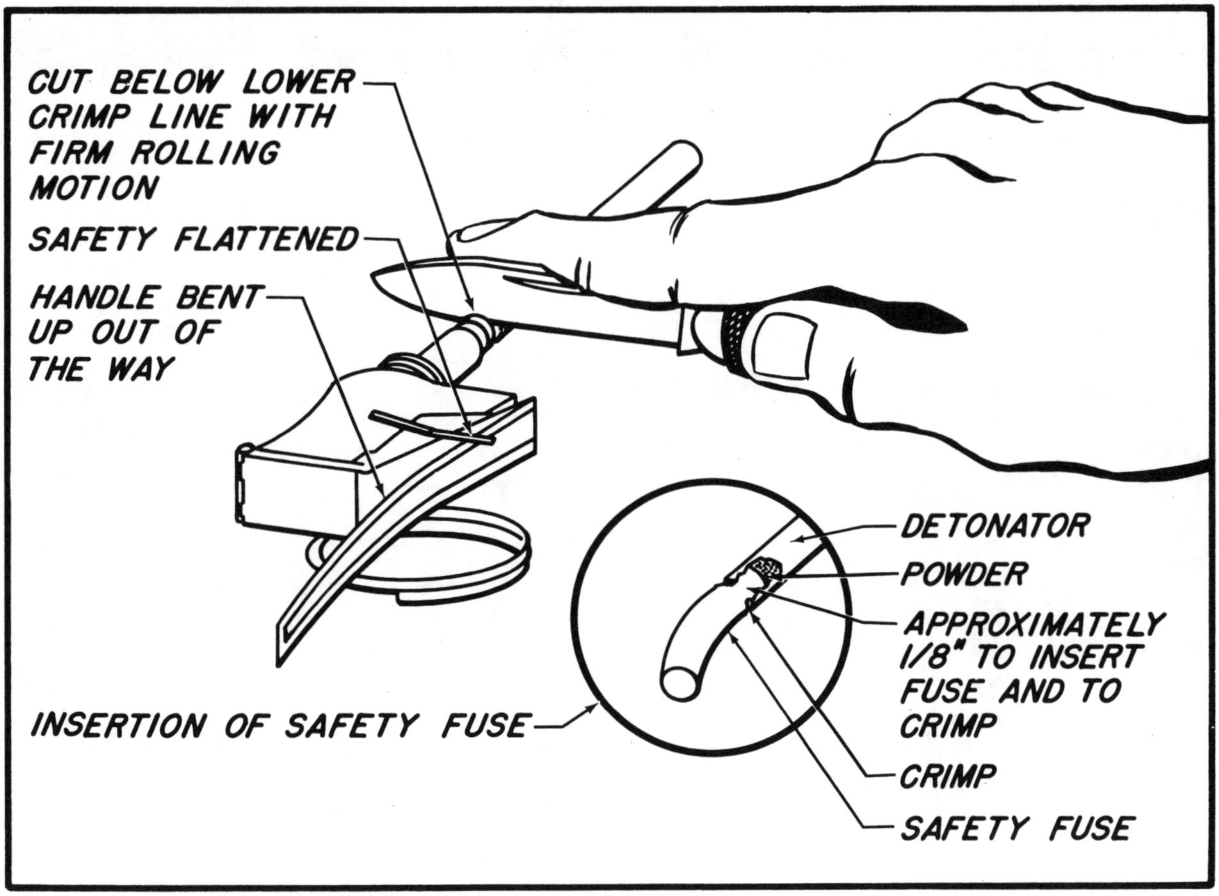

Fig. 22 — Removing Grenade Detonator

Now close the action of the rifle, release the safety, and retire to a safe position at least 50 yards away, being sure not to trip over the trip wire or pull cord. If the pull cord is used, take up any slack in the line and then pull firmly to fire the weapon and detonate the charge.

Frag Grenade Detonator

Nonelectric blasting caps can be obtained from fragmentation or other bursting grenades for use in detonating explosives. The detonator in a frag grenade is strong enough to set off pressed TNT, plastic explosives, and primacord. Detonators used in some countries are less powerful, but in all cases they will initiate primacord, which in turn can be used to set off any explosive. With weak detonators and no primacord, it may be possible to use two or three together to detonate less sensitive explosives such as plastics.

MATERIALS: Fragmentation grenade.

PREPARATION:

1. Spread and flatten out the points of the safety cotter pin to insure it will not be accidentally pulled out. Bend the fuse lever back out of the way.

2. The fuse assembly can now be safely removed from the body of the grenade with pliers. See Figure 21.

3. With a sharp knife cut around the detonator just below the lowest or second crimp mark where it is attached to the fuse assembly. This is most easily done by holding the knife firmly on the metal and slowly rolling the detonator-fuse assembly on a flat surface. See Figure 22. The soft metal of the detonator case cuts easily, allowing the detonator to be pulled off the nipple of the fuse assembly.

4. Inspection will reveal a shallow depression in the open end of the detonator. This depression, about 1/8 inch deep, is sufficient to permit the crimping of the detonator to time fuse if care is taken. See Figure 22. In lieu of crimping, the detonator can be taped to the fuse.

STEEL CUTTING TECHNIQUES

General

Many of the more critical target structures in sabotage operations are made of steel. Plastic explosives are the best for steel cutting. They are easier to place and permit the use of a simple rule of thumb for determining the amount of explosive needed rather than working out a complicated mathematical formula. Also, if the proper techniques are used much less explosive is required to do the job than demolitions formulas call for.

Rule of Thumb for Structural Steel

A strip of plastic explosive as thick as the steel to be cut, and as wide as it is thick, will cut through mild steel. This "ribbon" charge should be prepared in such a manner that it can be quickly placed and held in good contact against the steel where the cut is desired. See Figure 23.

When the exact dimensions of the target steel are known, a charge may be tailored in advance to fit perfectly. See Figure 24. If the target measurements can only be estimated, a more flexible charge can be prepared as shown in Figure 25.

MATERIALS: Wood scraps or tin strips, rags, string, blasting cap, plastic explosive.

PREPARATION:

1. Slice the explosive into strips of the same thickness and width as the thickness of the steel to be cut. This should be done with a sharp knife to preserve the density of the packaged explosive.

2. Wrap this explosive ribbon in one layer of cloth and tie it to a scrap of wood of the

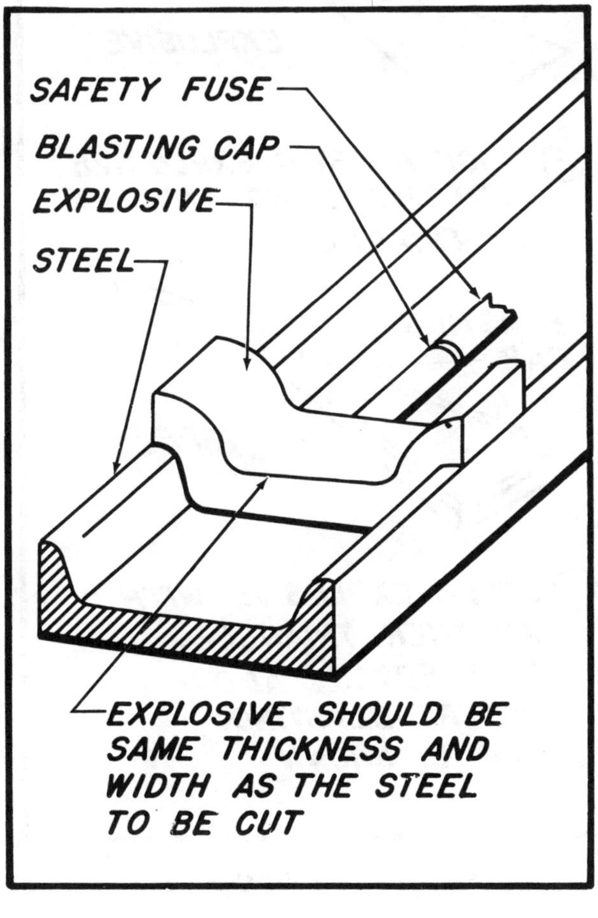

Fig. 23 — Ribbon Charge for Steel

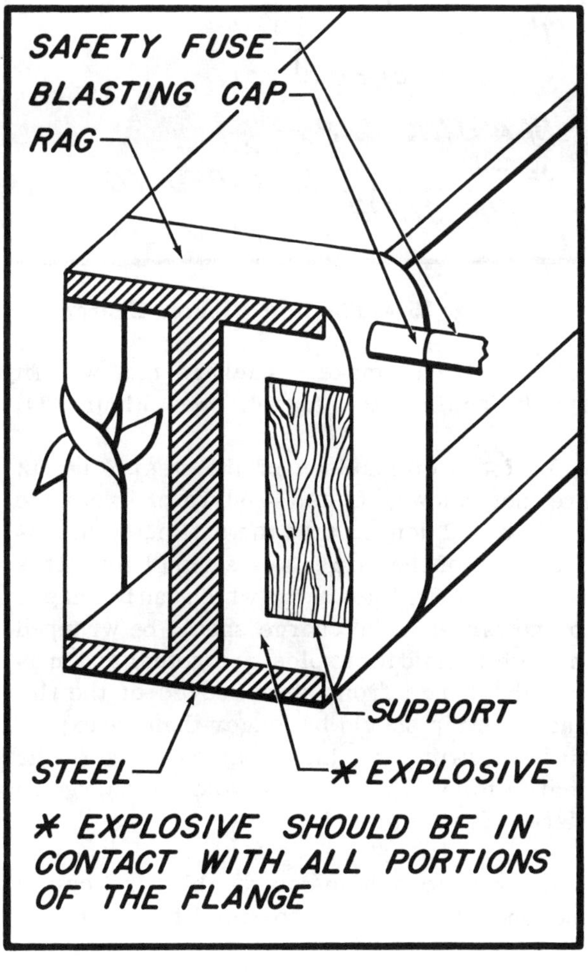

Fig. 24 — Tailored Ribbon Charge

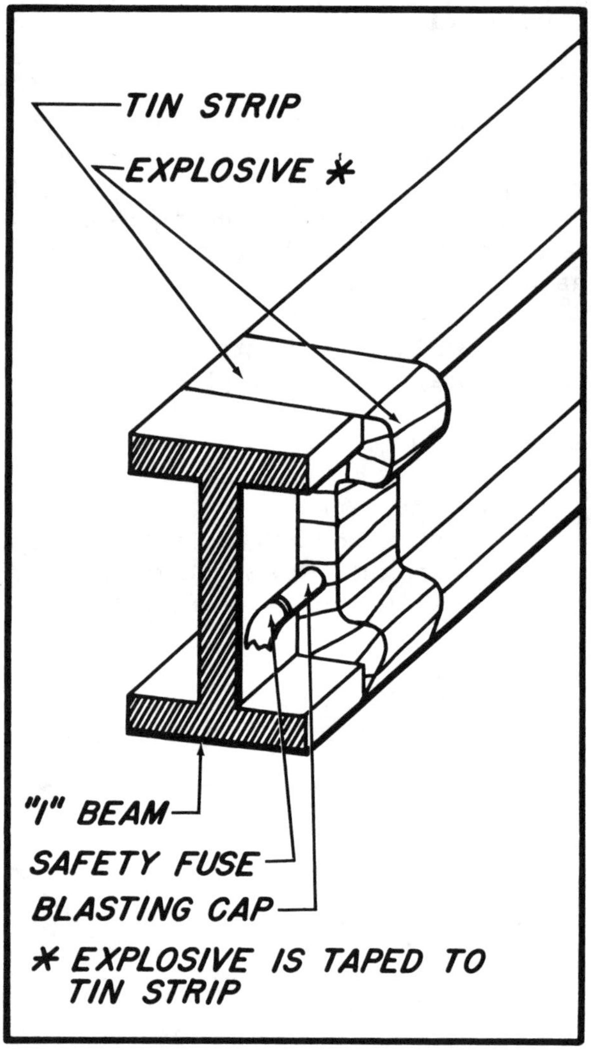

Fig. 25 — Flexible Ribbon Charge

proper size to make a package that will fit snugly against the target. See Figure 24.

3. If the dimensions of the steel to be cut are not known, fasten a strip of explosive thicker and longer than the estimated measurements of the target to a strip of tin. This provides a flexible charge which can be shaped to the target. The charge should be wrapped in cloth to hold the explosive together and protect the hands from sharp edges of the tin. The tin strip should be longer than the explosive, allowing 6 inches at each end of the charge for wrap-around to hold the charge in place. See Figure 25.

4. Prepare a hole for the blasting cap in one end of the charge to simplify priming at the target.

In placing the charge, fasten it securely on the target, being sure the explosive is in good contact with steel at all points. Insert the blasting cap and compress the explosive firmly around it.

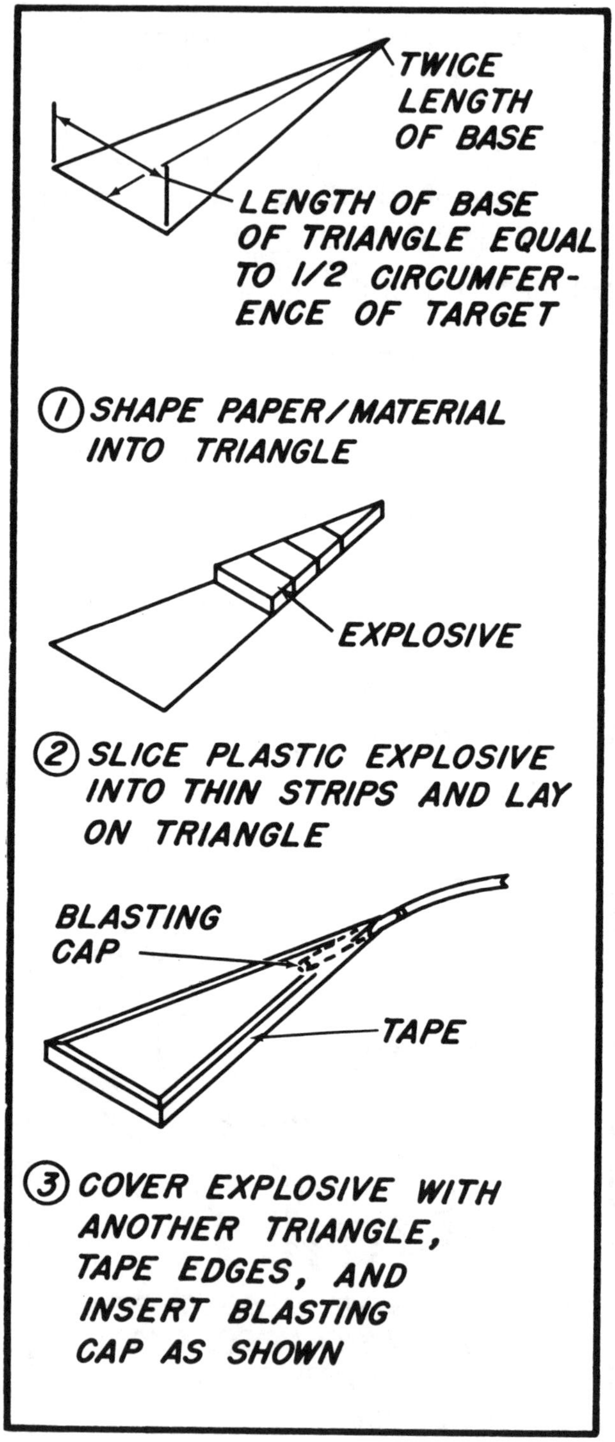

Fig. 26 — Saddle Charge

Cutting Steel Bars and Shafts

A triangular or "saddle" charge has been found to be highly effective in cutting bars and shafts. Its chief advantages are that it is easily placed and the cutting action is positive.

MATERIALS: Paper or cloth, tape, blasting cap, plastic explosive.

PREPARATION:

1. Prepare a paper or cloth pattern in the form of a triangle. The base of the triangle should measure 1/2 the circumference of the bar or shaft to be cut and the height of the triangle should be twice the measurement of the base.

2. Slice layers of plastic explosive 1/2 to 3/4 inch thick with a sharp knife. Make the slices as even in thickness as possible, being careful not to destroy the density of the explosive.

3. Place these layers of explosive over the triangular pattern until it is fully covered. Push the slices firmly together, leaving no air gaps in between, and trim off any edges extending beyond the pattern.

4. Cover the triangular charge with another piece of paper or cloth, wrap firmly and bind with tape to prevent the charge from falling apart. See Figure 26.

The triangular charge should be snugly wrapped, tied, or taped to the target bar or shaft, with the base of the triangle along the line where the cut is desired. The blasting cap should be inserted at the apex of the triangle and the plastic explosive pressed tightly around it.

STANDARD CHARGE

The standard charge provides a basic, versatile demolitions tool which can be made up in advance and utilized in many situations, especially when specific knowledge of the target is not available.

MATERIALS: Tape, primacord, plastic explosive.

PREPARATION:

1. Cut two pieces of primacord about a yard long. Onto each tie a triple roll or girth hitch knot with short pieces of primacord. The knots should be tight enough not to come untied, but they should slide easily along the main lines.

2. Split a half block of plastic explosive lengthwise and lay it open. Place the primacord lines and the two sliding knots between the two halves and adjust the knots so that one is just inside each end of the explosive. See Figure 27.

3. Close the two halves together over the primacord lines and knots. Tape or otherwise wrap and bind the charge tightly to insure close contact of explosive around the knots. Also tape the ends of the block so the knots will not pull out.

When completed, the charge can be slid along the primacord main line in either direction to permit adjustment on the target. Additional charges can be linked on by tying primacord ends together. Or, the single charge can be cut in two if smaller charges are desired and each will still have its own primacord leads and priming knot. If this is done, a simple knot should be tied in the ends of the primacord to keep the charge from sliding off.

Figure 28 suggests targets and placement of the standard charge.

NOTE: If primacord is in short supply, a single strand of primacord with two knots tied on it may be used in place of the dual system described and illustrated.

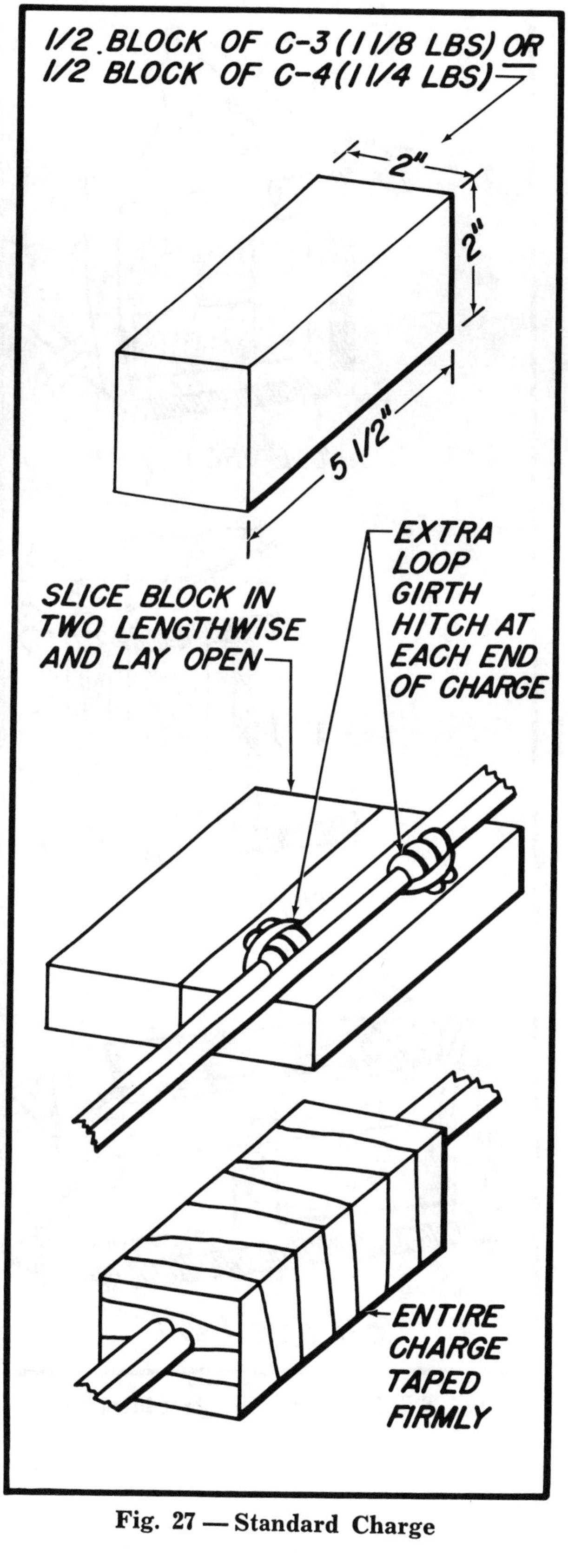

Fig. 27 — Standard Charge

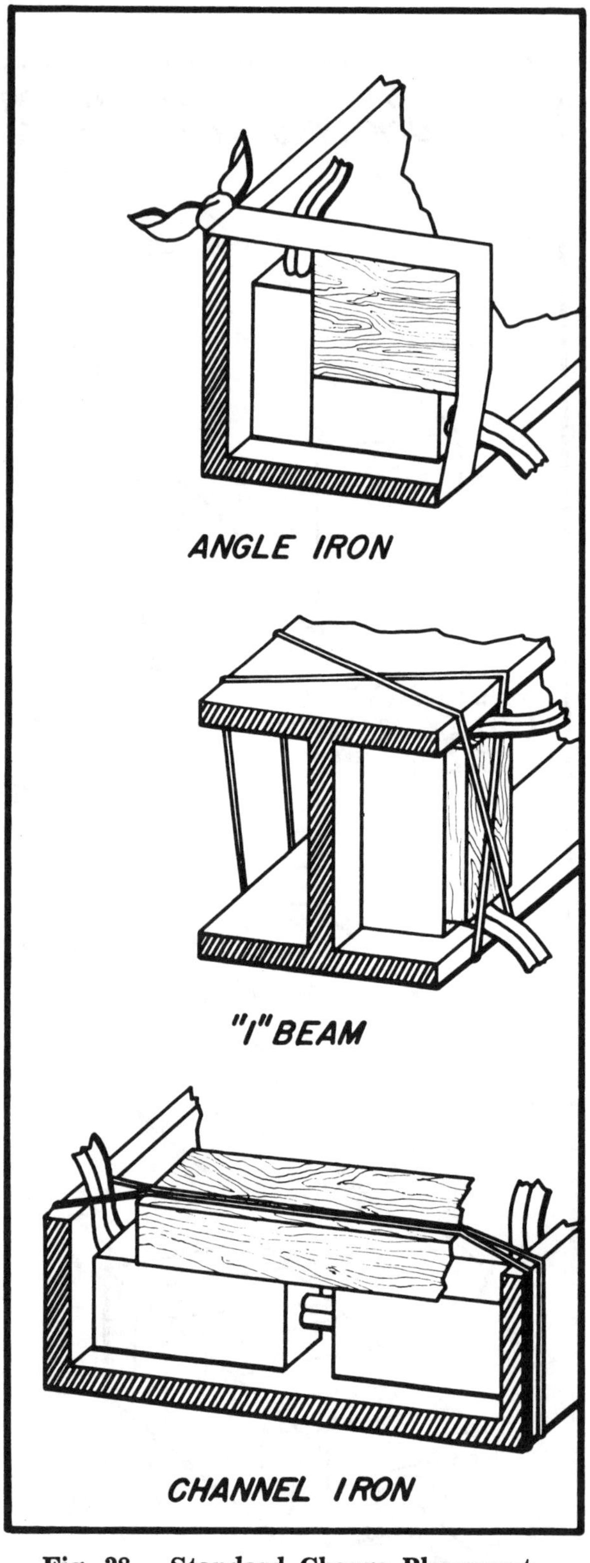

Fig. 28 — Standard Charge Placement

DERAILMENT TECHNIQUES

General

A primary resistance activity is damaging or destroying locomotives by derailment. To insure the derailment of a modern locomotive it is necessary to remove a length of rail equal to the length of the fixed wheel base of the locomotive. For practical purposes, removal of 20 feet of rail on one side of the track will result in the derailment of any locomotive.

The most effective place to make the cut is on the outside rail on a curve. Where two or more tracks parallel, derailment should be made in such a manner that a train wrecked on the inside track when rounding a curve will tear up and obstruct all the tracks.

The firing device used should be placed at least 5 feet ahead of the charges in the direction from which the train is expected, since the charges should detonate just ahead of the locomotive, not under it. If the direction of the train's approach is unknown, a firing device should be placed at each end of the string of charges.

This section describes two derailment techniques: one, a hasty method and, two, a concealed cratering system.

Hasty Derailment Method

When minimum time at the target is a primary consideration, the hasty method is an effective, quick way of taking out 20 feet of railroad track. The whole system can be prepared in advance and placed on target in a few seconds.

MATERIALS: Cord, string or rags, tape, primacord, plastic explosive, blasting caps, firing device.

PREPARATION:

1. Cut two lines of primacord at least 23 feet long and tie 8 to 10 triple roll or girth hitch knots at intervals of about 18 inches on each of them (see the previous section on the standard charge for instructions on tying the knots). The exact number of knots and the spacing will depend on the spacing of the crossties on the railroad track to be attacked. Be sure the knots are snug but loose enough to slide so adjustments in spacing of the charges can be made on the target.

2. Mold a half block of plastic explosive (minimum of 1 pound) around each pair of knots as described in the previous section on the standard charge and wrap or tape each

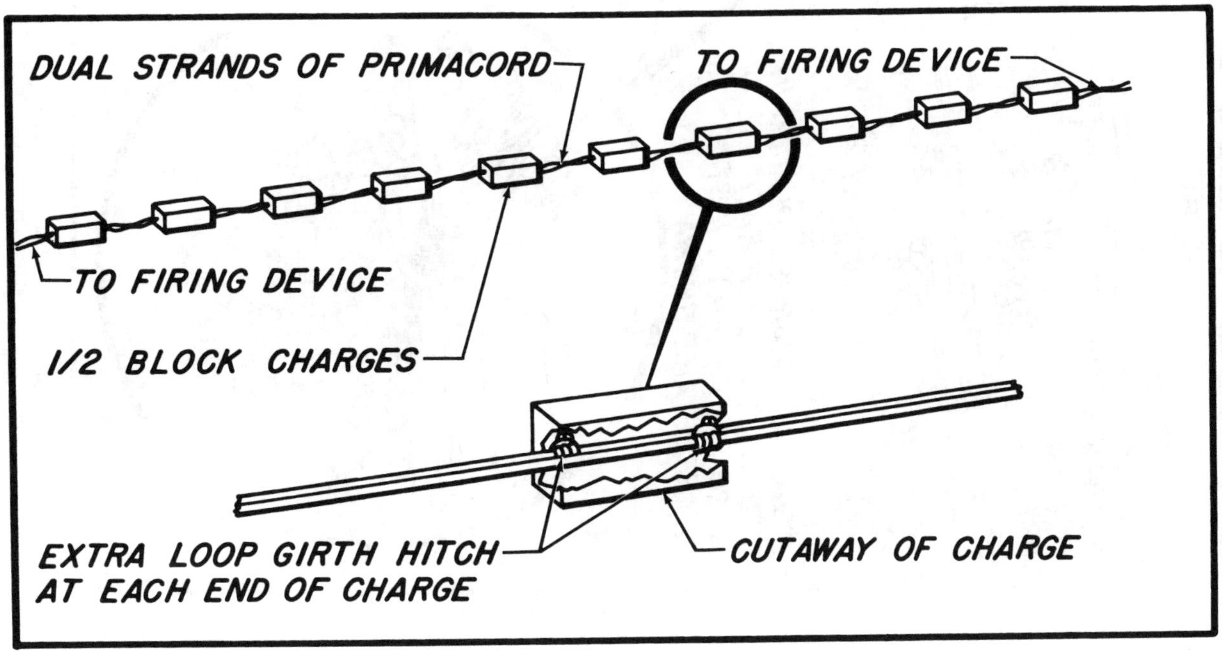

Fig. 29 — Hasty Derailment Charge

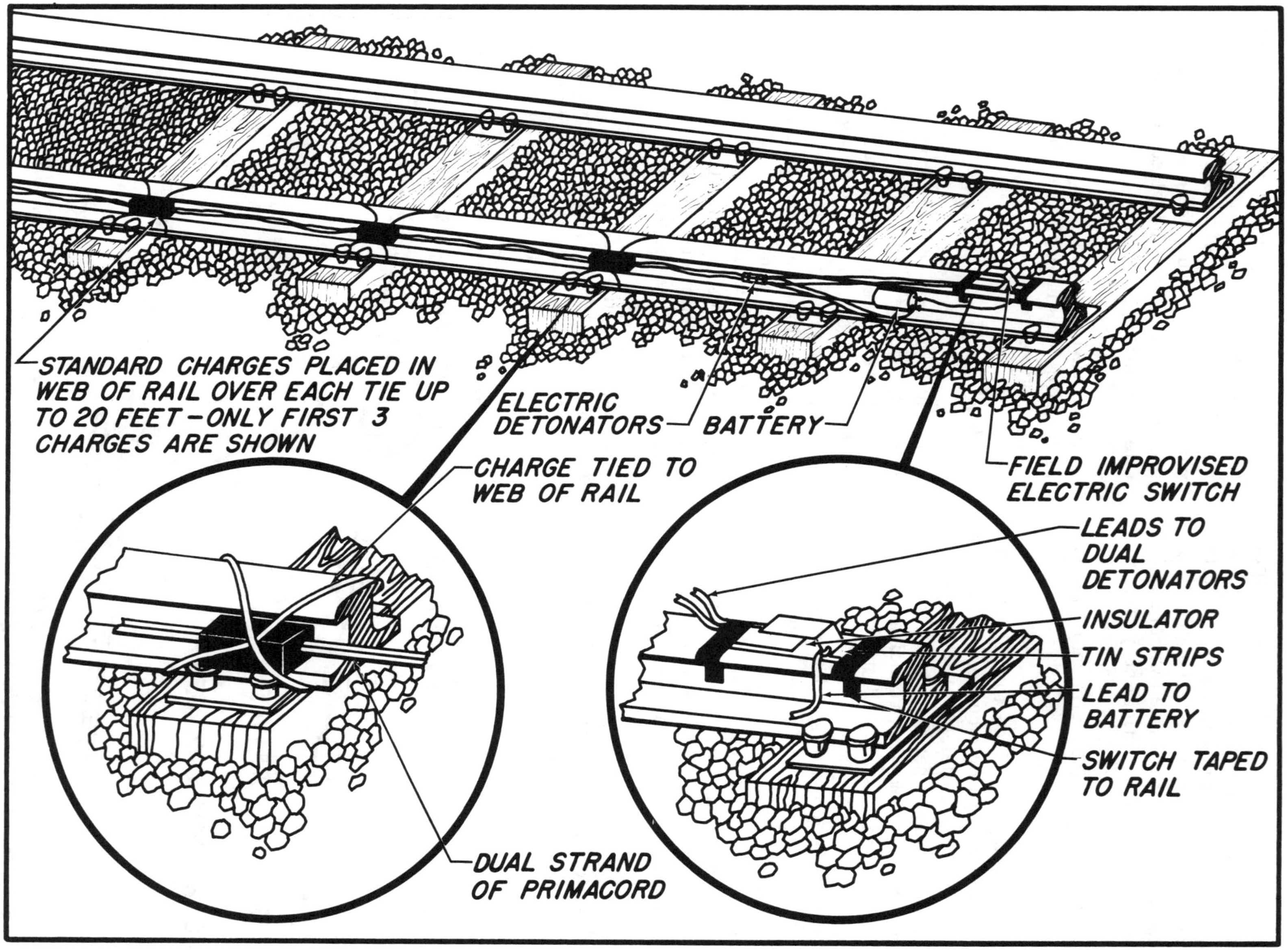

Fig. 30 — Hasty Method Placement

charge securely so it can withstand rough handling while being transported and placed on the target. See Figure 29 for details of construction. (NOTE: A single main line can be used if primacord is scarce.)

3. At the target, lay the primacord line and the attached charges straight out along the section of rail to be cut. Be sure there are no kinks or sharp bends in the line. Attach a charge to the web of the rail directly over each tie in the 20 foot section. If strong magnets are available they can be fixed to the charges and provide a very quick means of attachment. Otherwise use cord, strips of cloth, wire, or tape. See Figure 30.

The special railroad torpedo firing device is the best initiator to use with this hasty method. It would be clamped to the top of the rail and attached to the chain of charges with a 5 or 6 foot length of primacord.

Among the alternative methods of initiation which could be prepared in advance and used in this technique is the electric pressure switch improvised out of metal straps which was shown in Figure 7. Figure 30 illustrates this application. BE SURE, however, that the metal straps are completely insulated from the rail by a sheet of rubber, wood, cardboard, or heavy paper.

Concealed Cratering System

If the operational situation permits extra time on the target and some noise, the concealed cratering technique for derailment is more sure-fire since it also tears up ties and roadbed. It also alleviates the danger of detection by track inspection prior to the passage of a train.

MATERIALS: Tape, primacord, plastic explosive, blasting caps, firing device.

PREPARATION:

1. Cut two main lines of primacord about 23 feet long or longer and 4 or 5 shorter pieces to serve as branch lines. (NOTE: A single main line can be used if primacord is scarce.)

2. Tie two sliding triple roll or girth hitch knots (see section on standard charge) to each of the branch lines.

3. For each branch line slice open a FULL BLOCK of plastic explosive lengthwise, and mold it around the middle of the branch line with a knot just inside each end of the block.

4. Cut the block of explosive crosswise into two equal lengths, being sure not to cut through the primacord in the center. Now fold the two portions of the block together (see Figure 31) and tape or tie securely. Tie or tape the free ends of the primacord together. This method of constructing the charge will permit much rougher handling in transporting it and placing it on the target than if a single strand of primacord with a single knot in the end is used.

5. To install this system at the target, scoop out ballast as necessary and bury one charge IN EVERY OTHER SPACE BETWEEN TIES directly under the rail to be removed. Attach the branch lines from the charges to the main primacord line and cover the whole system with ballast. See Figure 31.

A pressure firing device which can be concealed under the rail or a tie is best to use with the cratering system. Figure 19 illustrates such an improvised pressure device, utilizing a modified M-2 fuse lighter.

IMPROVISED CRATERING EXPLOSIVE

In general, improvised explosives are too complicated to prepare or too weak for steel cutting and most other sabotage tasks. However, if the problem is to get a quantity of lower velocity explosive for cratering or moving a large volume of earth or rock, a good simple one can be prepared from ammonium nitrate fertilizer.

This improvised explosive can be used for attacking bridge abutments, or causing a landslide to block a mountain pass, close a tunnel, derail a train, or similar purposes.

To prepare the explosive, simply mix about one gallon of fuel oil or motor oil to one bag (80-100 pounds) of fertilizer. The mixing can be done in the bag or in the bore hole prepared to receive the charge. Use about the same amount of the mixture in weight as would be required of dynamite.

Best results are obtained with ammonium nitrate fertilizer which has a 33 percent nitrogen content. For reliable detonation and maximum cratering effect the mixture should be in a confined space such as a bore hole and tamped with earth or otherwise covered and compressed. Also it is easily weakened by moisture and should not be used in a wet situation unless it is sealed in a watertight container.

Because this improvised explosive is extremely insensitive, a booster is required to initiate it. A half pound charge of any high explosive in the bottom of the hole will be sufficient to insure detonation.

While insensitive to shock, the mixture is a fire hazard and normal precautions against fire should be observed. If 50 pounds or more is being used in one charge, anyone within a few hundred yards is in danger of being struck by flying debris.

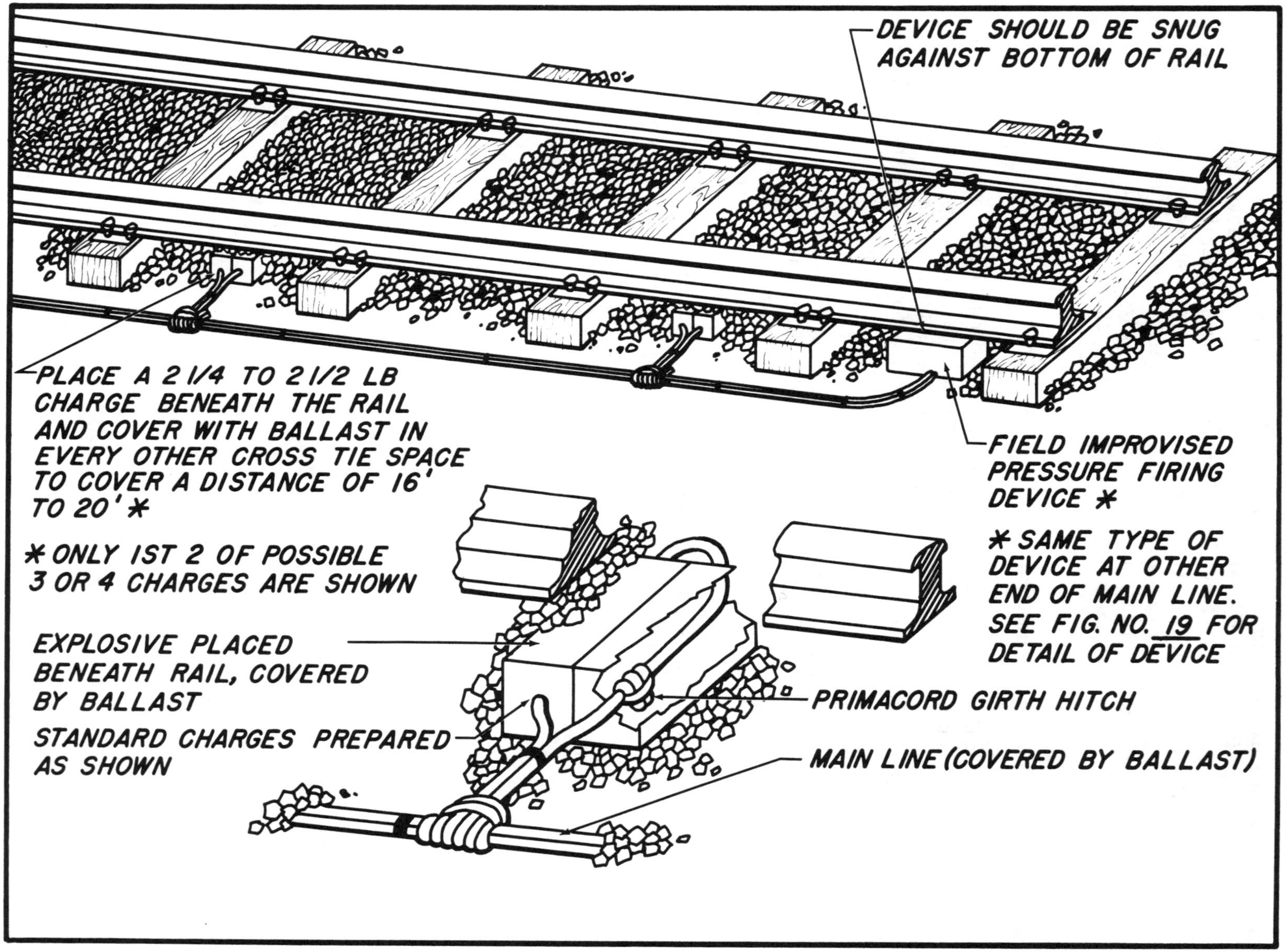

Fig. 31 — Concealed Derailment System

CONE SHAPED CHARGES

General

Shock waves from a detonating block of explosive emerge in all directions, creating a generalized blast effect. It is possible, however, to shape an explosive charge in such a way that a portion of the shock waves will be focused and concentrated on a very small area, thereby creating a very powerful puncturing or drilling action.

A familiar military use of this principle is the 3.5 inch antitank rocket, or bazooka round. The explosive charge in the head of the rocket has a conical cavity in the forward end. When the charge is detonated against the side of a tank, the shock waves emerging from the conical cavity are focused on one spot where they will penetrate several inches of tough steel.

The cone shaped charge has excellent possibilities for sabotage especially for use against industrial machinery. The 3.5 rocket itself provides one of the very best shaped charges. A later section on this versatile weapon will describe and illustrate how the rocket head can be removed and used as a placed shaped charge.

Lacking a 3.5 round, an effective shaped charge can be fashioned around any cone shaped glass object. Following are directions on how to do it with a martini glass or a wine bottle with a dome shaped bottom.

Martini Glass Shaped Charge

MATERIALS: Martini or wine glass, beer can or other tubular container or cardboard from which to fashion one, sticks, tape or string, plastic explosive.

PREPARATION:

1. File a notch around the stem at the base of a martini or wine glass as shown in Figure 32. Rap the stem sharply against a solid object to break it off from the cone of the glass.

2. Place the glass upside down in any cylindrical container or cardboard tube fashioned to the diameter of the mouth of the glass. Unless the glass is wedged firmly in the container, tape it around the rim to hold it in place.

3. Pack plastic explosive behind the apex

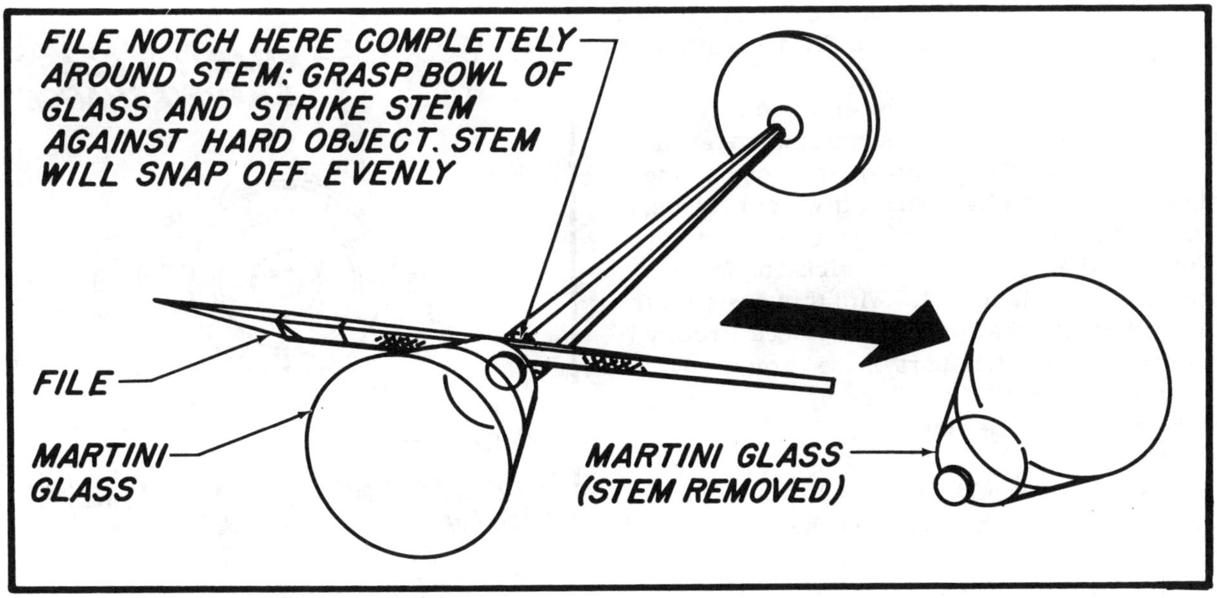

Fig. 32 — Removing Glass Stem

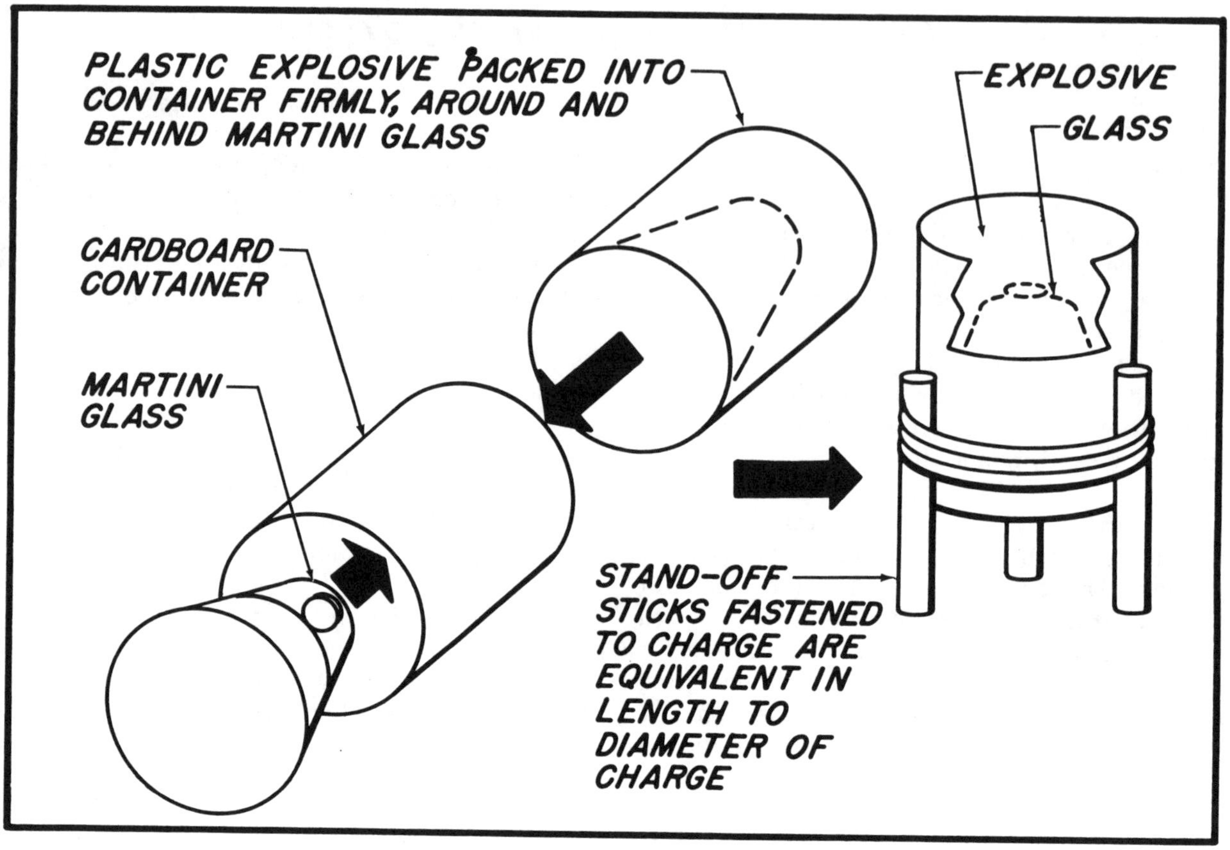

Fig. 33 — Martini Glass Charge

of the glass, carefully tamping it with a stick to provide good density and avoid air pockets. For best results the explosive should be packed to a height above the apex of the cone equal to the height of the cone itself. To conserve explosive, the top of the charge may be dome shaped.

4. Tape, wrap, or bind the whole charge to insure that it remains in a firm condition.

5. This charge should be elevated from the target approximately the same distance as the diameter of the mouth of the cone so the shock waves will have room in which to focus. This stand-off can be provided by taping or tying to the charge three sticks to form a tripod. See Figure 33. With this system of placement, the charge would be placed directly over the most vital part of the target. With a vertical target, a counterweight system of placement could be employed. See Figure 34.

The charge should be primed with a blasting cap placed in the center of the rear of the charge. Do not insert the cap so far that it touches the top of the glass cavity. Tape or tie the cap in a firm upright position so it will not become dislodged.

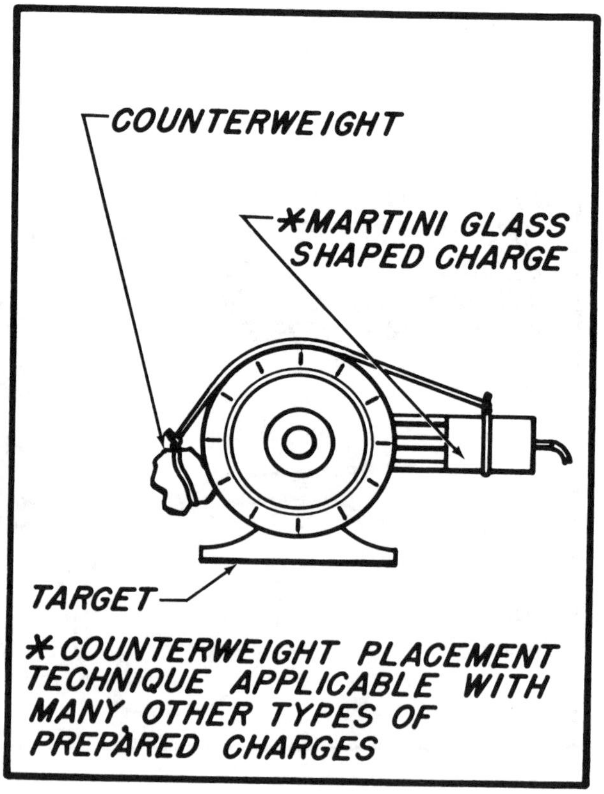

Fig. 34 — Counterweight Placement

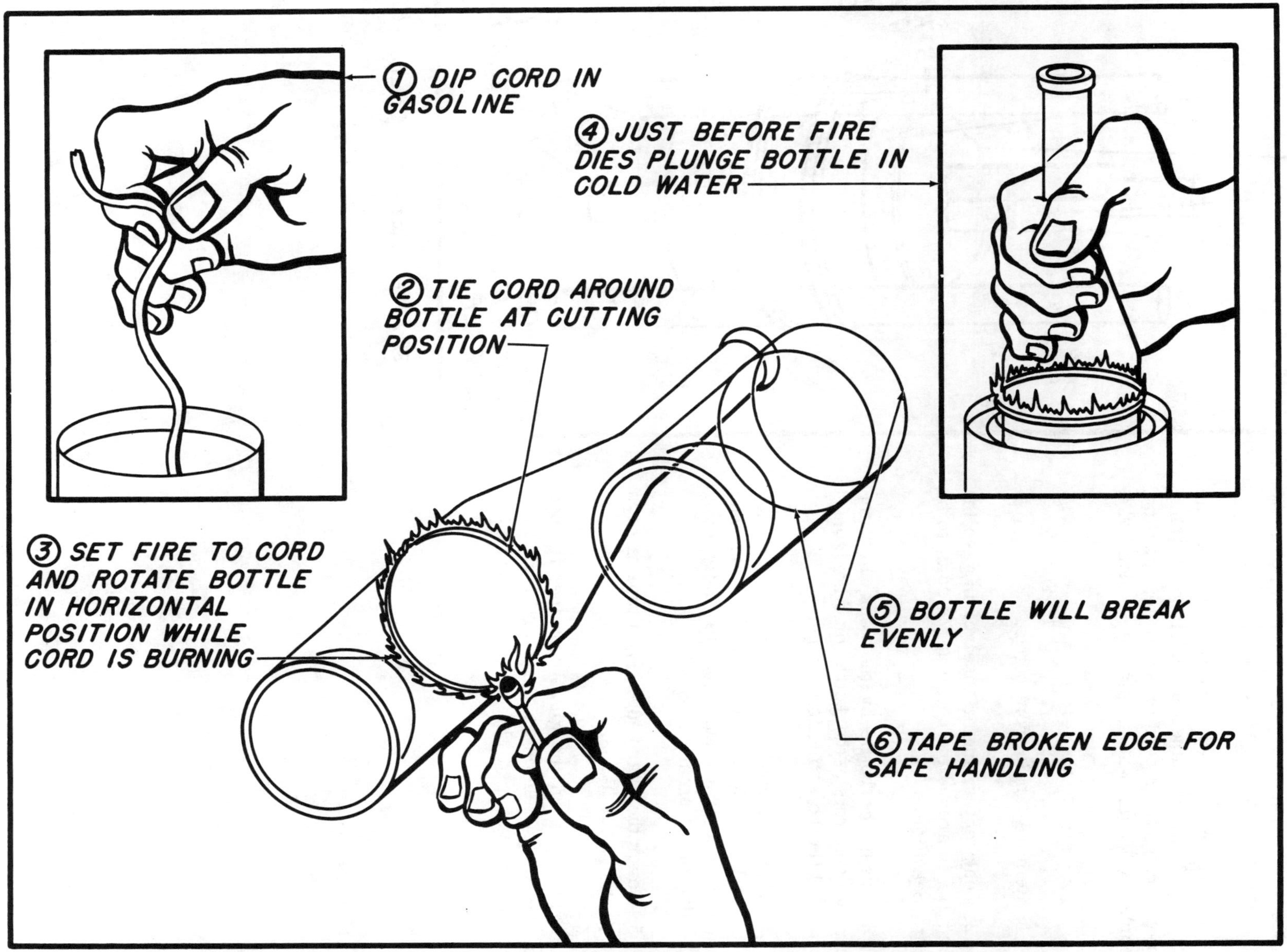

Fig. 35 — Cutting Wine Bottle

Wine Bottle Shaped Charge

The bottom portion of a wine bottle which has a dome or conical shaped cavity can be used in the same way as a martini glass to construct a cone shaped charge.

MATERIALS: Wine bottle with a cavity base, tubular container, sticks, tape or string, mop string, fiber cord or shoelace, gasoline, bucket of cold water, plastic explosive.

PREPARATION:

1. Remove labels from bottle.

2. Soak mop string or shoelace in gasoline and tie it around the bottle at the point to be cut. See Figure 35.

3. Grasp the bottle by the neck, light the string, and rotate the bottle on a horizontal plane while the string is burning.

4. When the burning starts to die out plunge the bottle into a bucket of cold water. The bottle should crack and separate easily along the line of the string.

5. Cover the sharp edges with tape or rags to protect your hands.

From here on the glass container is used in the same fashion as the martini glass to construct a shaped charge. See Figure 36.

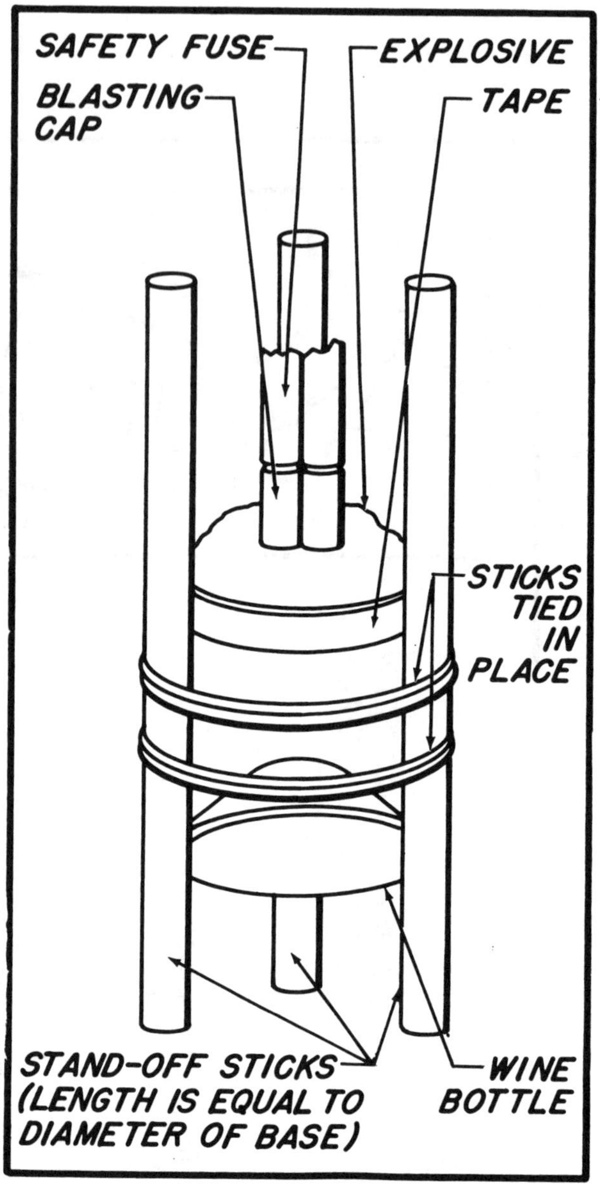

Fig. 36 — Wine Bottle Charge

PLATTER CHARGE

The platter charge technique is a good one to use in situations where it is possible to get near the target but not possible to place a charge directly on it. The charge uses high explosive to hurl a heavy piece of steel at great velocity. It is effective against such targets as transformers, generators, fuel storage containers, and vehicles. It can be fired horizontally or used as an improvised mine.

At close range the platter will penetrate about 1 inch of mild steel. Or it will penetrate woven steel (chain link) fencing and continue on with enough velocity to do severe damage to a target beyond. It is effective at ranges up to 100 feet or more, although at this distance sighting becomes a problem unless the target is a very large one.

MATERIALS: Paper, tape, cloth, tin or cardboard container, steel plate, primacord, blasting cap, plastic explosive.

PREPARATION:

1. Select a piece of steel plate at least 1/4 inch thick and 4 inches or more on a side. A 3 to 5 pound plate is about right.

2. Pack plastic explosive densely against one side of the plate, completely covering it. Use about the same amount of explosive as the weight of the plate.

3. Secure the explosive firmly to the plate with cloth, paper, tape, or string. A firmer charge can be made if a cardboard or metal container is fashioned to hold the explosive in shape behind the plate.

The charge may be primed with a blasting cap or a primacord triple roll knot. The cap or knot should be embedded in the center of the explosive behind the plate but not pushed in so far that it touches the plate. See Figure 37.

If the charge is to be fired horizontally at a target it should be solidly supported above the ground on a box or tied to a stake or tree and "bore sighted" with the plate side toward the target. See Figure 38.

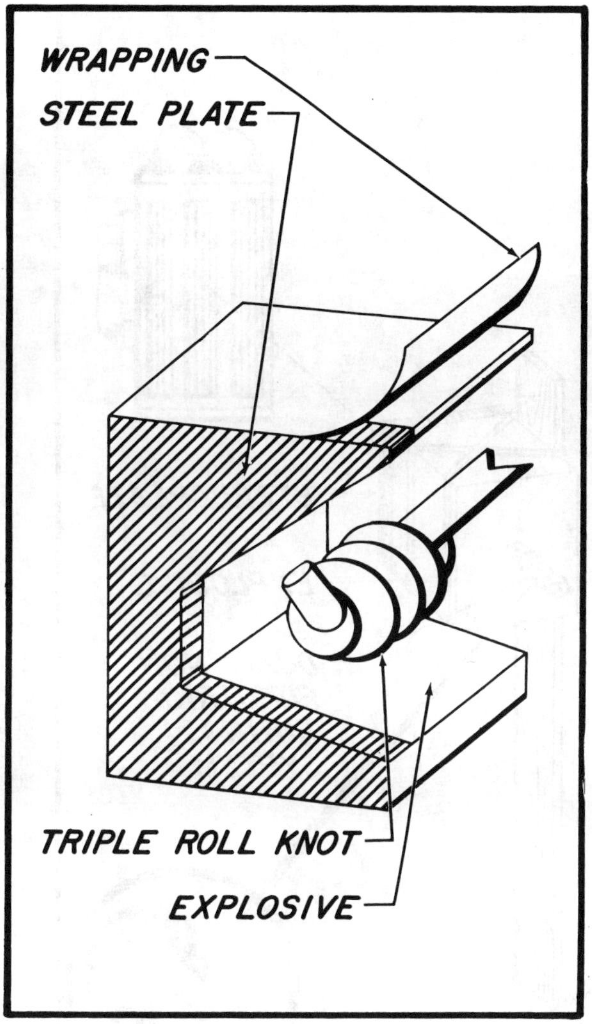

Fig. 37 — Platter Charge

When used as an improvised mine, the charge should be buried just under the surface of the ground, plate side up. It should be primed with a primacord lead to an appro-be primed with a primacord lead to an appropriate firing device. Placement should be such that when the firing device is initiated the charge will be detonated under the engine, passenger compartment, or the fuel tank of the target vehicle. See Figure 39.

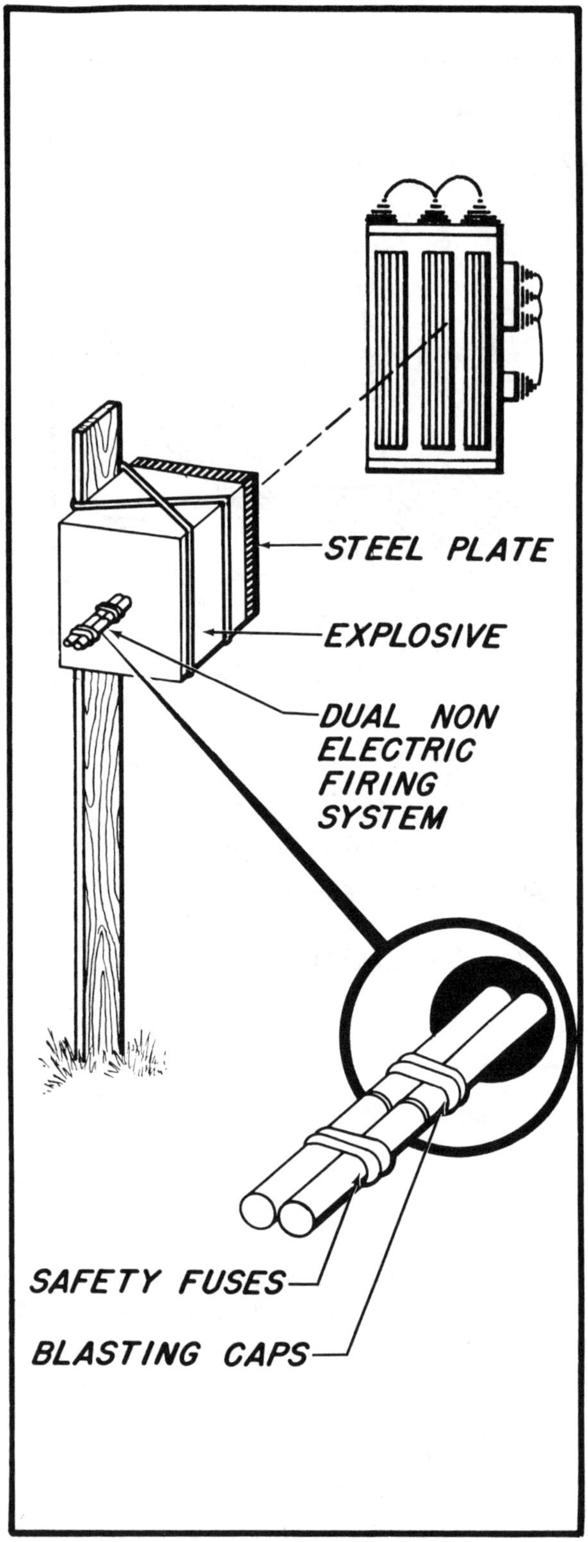

Fig. 38 — Platter Charge Placement

Fig. 39 — Platter Charge Mine

COUNTERFORCE CHARGE

The counterforce effect is obtained by placing two identical (earmuff) charges on exact opposite sides of a target and detonating them simultaneously. The detonation produces a collision of shock waves within the target with severe destructive effect.

The main value of the counterforce technique is that it requires only half the explosive needed to accomplish the same result using standard formulas and placement. For example, a 14 inch diameter timber which requires 3-3/4 pounds of plastic explosive to cut conventionally can be sheared in two with counterforce charges of 3/4 pound each. Reinforced concrete piers 2 feet by 2 feet which can be broken conventionally by 17 pounds of explosive can be broken by counterforce charges of 4 pounds each.

Counterforce charges are most effective against targets which contain no air spaces and which have a high degree of consistency in internal composition, so that the opposing detonation waves encounter the same substances on their way to the midpoint of the target. Solid wood and concrete targets are more vulnerable to the shattering effect of the counterforce charge than are most metal targets.

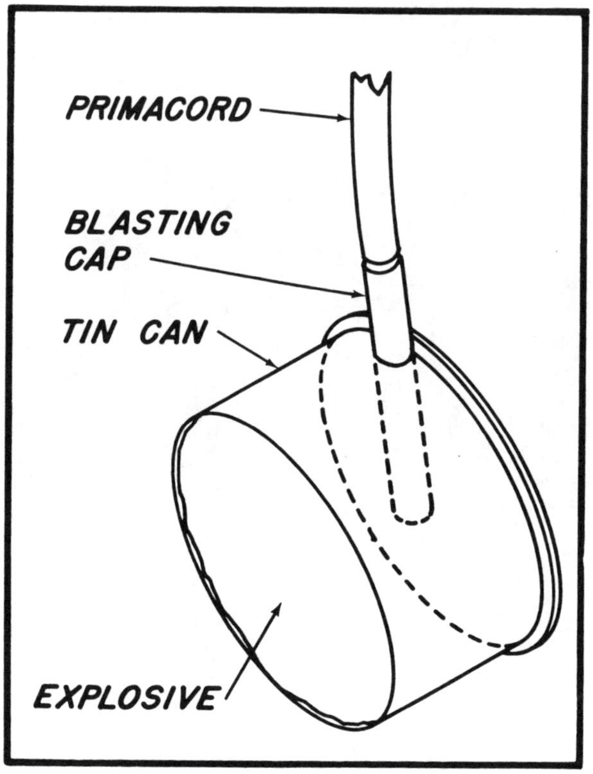

Fig. 40 —Counterforce Charge

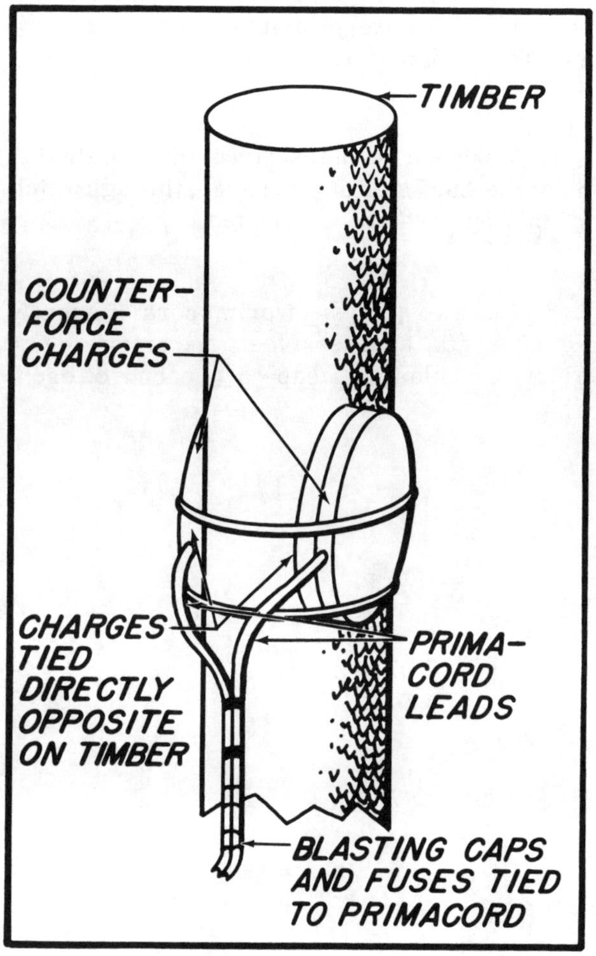

Fig. 41 — Counterforce Charge Placement

The counterforce technique requires care in preparation and placement on the target. It should, therefore, be restricted to situations where the explosive saving is necessary and conditions permit the additional time on the target required.

MATERIALS: Tape, cord, containers for explosive, primacord, blasting caps, plastic explosive.

PREPARATION:

1. Pack plastic explosive densely into two

identical containers -- such as jar lids with an adequate lip or the bottom portions of tin cans. The size of the containers and the amount of explosive will depend upon the type and size of the target. Potency of the charge is increased more, proportionately, by widening its diameter than by making it thicker.

2. Make a hole in the rim of the containers, near the back side of the charge, through which a blasting cap can be inserted.

3. Cut two pieces of primacord of exactly equal length, at least 6 feet, and crimp a non-electric blasting cap to one end of each. Tape or tie the other ends of the primacord together.

Priming is best done at the target. Insert a blasting cap through the hole in the rim of each charge until the business end of the cap is in the center of the charge. See Figure 40.

Place the charges on exact opposite sides of the target. This may be done by taping or tying them into place (see Figure 41) or by suspending them over the target on a string and permitting them to counterbalance each other.

Attach a suitable firing system to the taped ends of the primacord.

DIAMOND CHARGE

The diamond charge technique utilizes colliding shock waves within an explosive charge to obtain a cutting or shearing effect. While it may be used to cut through any material, it is most effective against steel, especially cylindrical steel members such as shafts. It must be carefully prepared and placed but it does permit a considerable saving in explosive over standard charges and placement.

MATERIALS: Paper or cloth, tape, blasting caps, primacord, plastic explosive.

PREPARATION:

1. Construct a diamond shaped pattern on paper or cloth. The long axis of the diamond should be the same length as the circumference of the target -- plus a couple of extra inches to make up for "loss" of length when the charge is wrapped around the target. (NOTE: If you are up on your math you can allow an extra inch for every inch of the diameter of the target.) The short axis of the diamond should be half the length of the long axis.

2. Slice layers of plastic explosive at least 1/4 inch thick with a sharp knife. Make the slices as even in thickness as possible, being careful not to destroy the density of the explosive.

3. Place these layers of explosive over the diamond pattern until it is fully covered. Push the slices firmly together, leaving no air gaps in between, and trim off any edges extending beyond the pattern.

4. Cover the charge with another piece of paper or cloth, fold over the edges and tape tightly to retain the charge in a flat diamond shape. See Figure 42.

5. Cut two pieces of primacord of exactly equal length, crimp a nonelectric blasting cap to one end of each and bind the other ends together with string or tape.

To use on a steel shaft or bar, wrap the charge completely around the target so the two points of long axis meet, and tape or tie it snugly in position. See Figure 43. The points should just meet, but a slight gap or overlap is not critical. The cutting action of the charge will be along the line of the long axis of the charge.

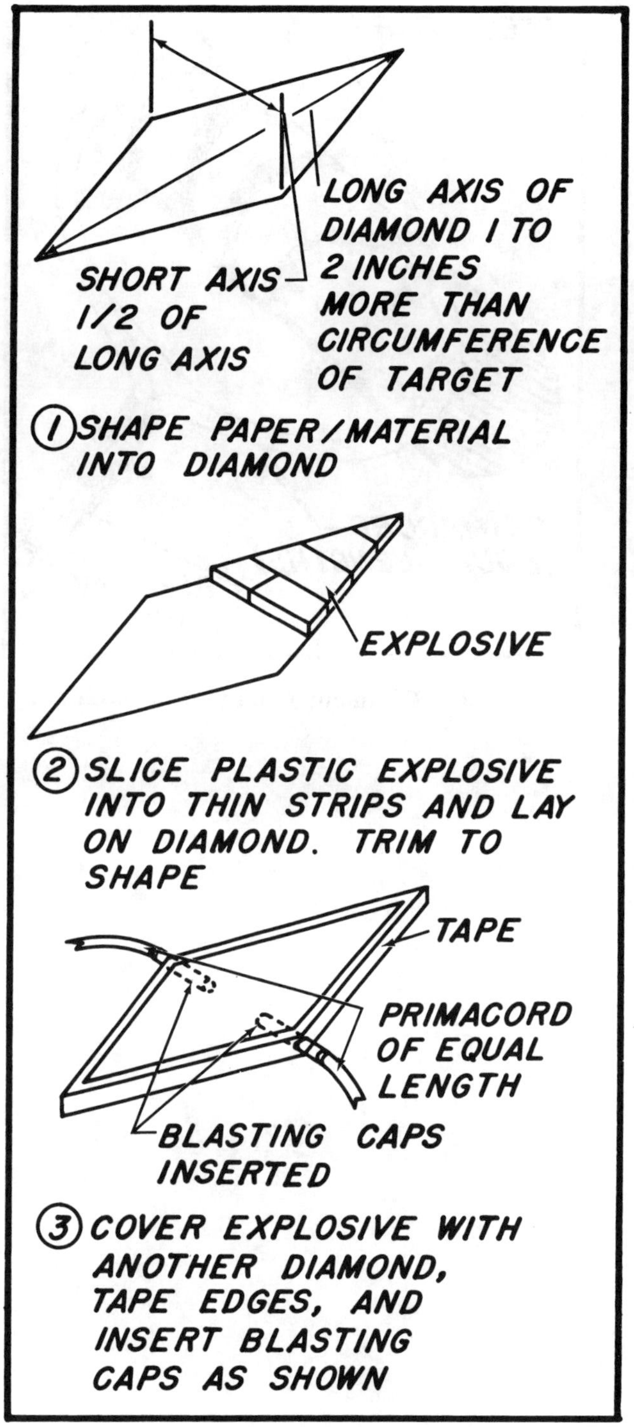

Fig. 42 — Diamond Charge

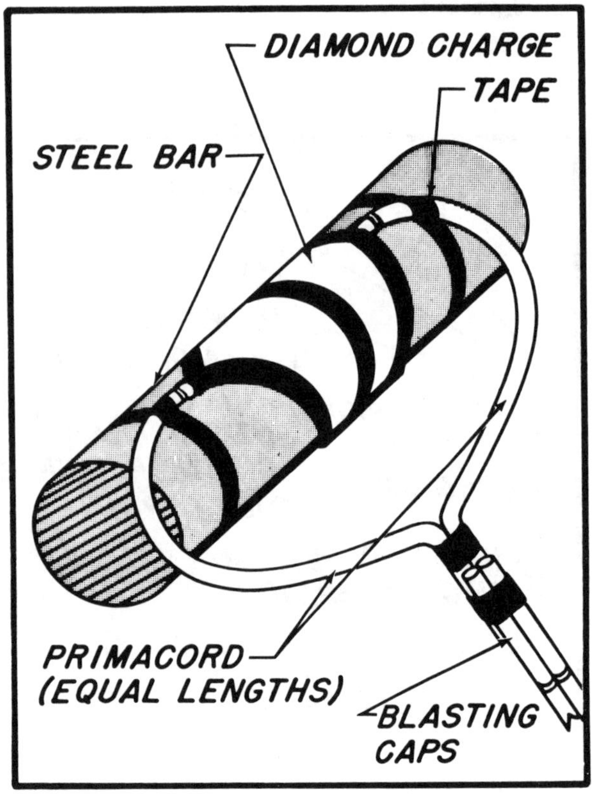

Fig. 43 — Diamond Charge Placement

To prime, insert a blasting cap into each point of the SHORT axis. Be sure each cap is embedded to an equal depth in the explosive and is secured firmly in place.

Attach a suitable firing system to the taped ends of the primacord.

CABLE CUTTER

This technique provides a way of cutting overhead communitions or power cable with explosives placed from the ground and not easily removed if discovered.

MATERIALS: Tin can or bottle, sand, stones, strong cord, TNT or plastic explosive, nonelectric blasting cap, delay firing device (time pencil or time fuse).

PREPARATION:

1. Prepare a charge of TNT or plastic explosive 1/2 to 1 pound in weight.

2. Prepare a counterweight approximately 1-1/2 times the weight of the charge. The counterweight should be longer than the charge and cylindrical in shape to permit it to slide or roll over the target cable easily. A tin can filled with rocks, a bottle filled with sand, or a section of rod or cable will work very nicely.

3. Lay the charge and the counterweight parallel to each other about a foot apart and tie the top and bottom ends of the charge to the top and bottom of the counterweight with strong cord and non-slip knots. See Figure 44.

4. Attach a wire loop securely to each end of the counterweight. Grenade fuse pull rings are ideal, or the rings can be fashioned by bending nails or stiff wire into loops.

5. Cut a throwing line of fairly heavy smooth cord about four times as long as the distance from the ground to the target cable. The suspension lines of a parachute are excellent if available.

6. Pass one end of the throwing line through the two loops on the counterweight and pull it through until the ends of the line are equal in length. To these free ends securely attach a throwing weight such as a rock or chunk of metal.

7. Leaving the charge and counterweight on the ground underneath the target cable, and the excess of the throwing line neatly coiled so the line will run out easily, pitch the throwing weight over the cable and allow it to descend to within reach. See Figure 44.

8. Now prime the charge with a nonelectric blasting cap and a delay device (time pencil or safety fuse) and initiate the delay. Dual priming is advisable, particularly if the time pencil delay is used.

9. Remove the throwing weight from the ends of the throwing line, and hoist the counterweight and the charge up to the cable. Ease the counterweight over the cable and down until the charge is snug up against the cable. See Figure 44.

10. Separate the ends of the throwing line and by pulling gently on one end ease the line through the rings on the counterweight and retrieve it. See Figure 44. The charge and counterweight are left suspended on the cable and the throwing line can be used for placement of other charges along the cable if desired.

PRECAUTIONS:

a. Never use a conducting material such as wire for a throwing line.

b. Never attack high power lines by this method in wet weather, because a wet throwing line could act as a conductor.

c. Be sure to allow enough delay time after the charge is placed for the operator to retire to a safe distance. When cut, the cable may whip around in any direction.

d. While not too pertinent for this technique since the operator remains on the ground, a good rule to observe with any high voltage equipment is never approach it closer than the length of the largest visible insulator used on the equipment.

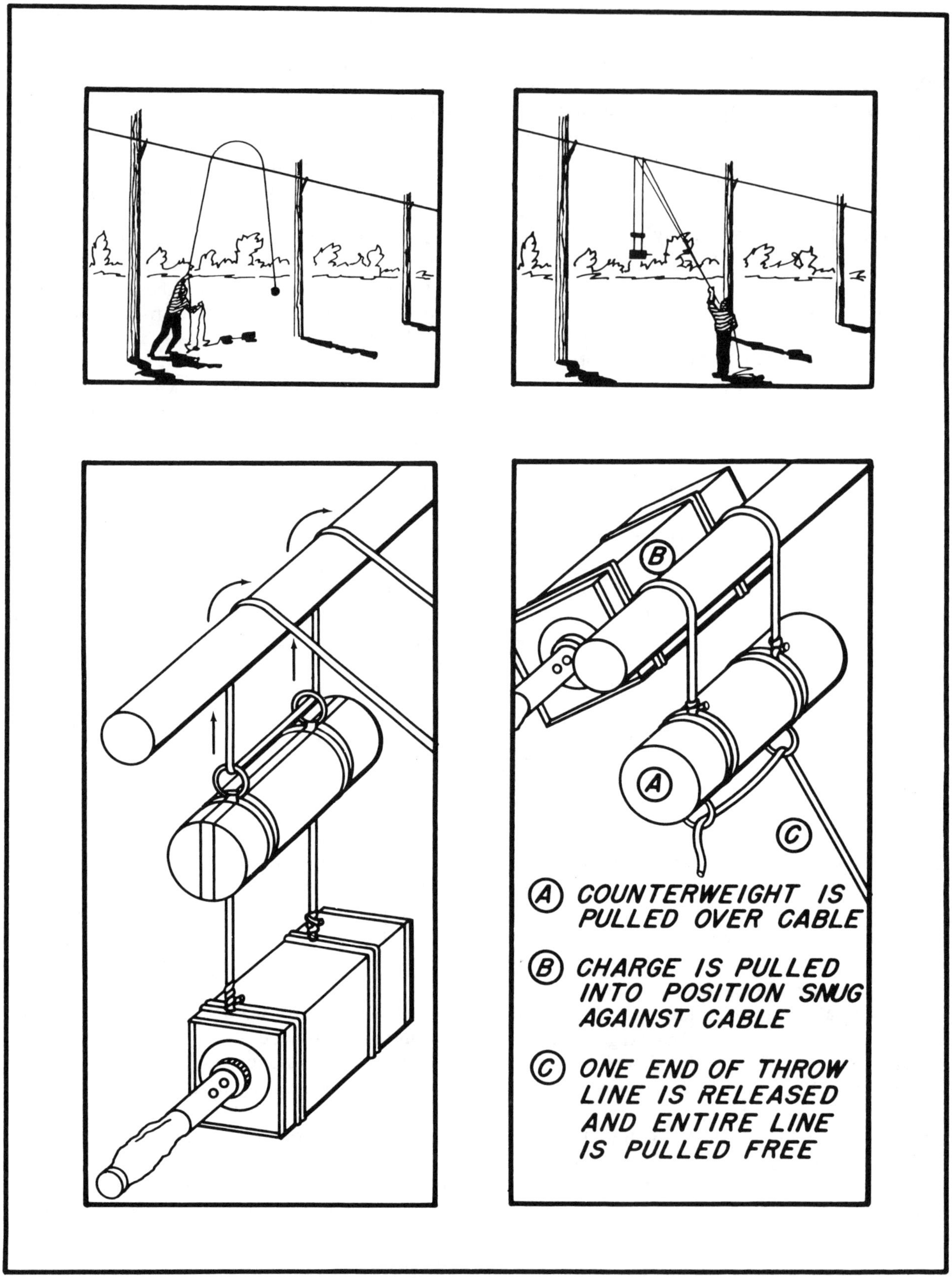

Fig. 44 — Cable Cutter

AMBUSH OR ANTIPERSONNEL CHARGES

Scatter Charge

This charge is very effective as an ambush weapon against personnel on foot or in unarmored vehicles. It also may be used against communications equipment such as radar consoles or telephone switchboards.

When detonated the charge throws a heavy pattern of metal fragments toward the target at which aimed. The range is not great -- perhaps 50 yards maximum -- but the fragments disperse rapidly for broad coverage at ranges of 10 to 20 yards. The effect is devastating to both equipment and morale. Fragments will penetrate metal of an unarmored automobile or truck, corrugated sheet metal or light wooden walls such as those of a mess hall or barracks.

MATERIALS: Metal fragments (nails, washers, bolts, nuts, etc.), rags, wire, tape, or string, container, primacord and/or blasting caps, plastic explosive or dynamite.

PREPARATION:

1. Pack explosive densely into the bottom one fourth of the container. A tin can of about 1 gallon capacity is about ideal for most targets. The container should not be smaller than 1 quart but large diameter cardboard tubes or ordinary pails are satisfactory.

2. Puncture a hole in the bottom center of the container large enough for the insertion of a blasting cap or primacord. If initiation is to be by primacord, the primacord should be inserted through the hole and a triple roll knot tied inside the container. The explosive should then be packed around the knot.

3. If the charge is not to be used in the near future or may be subjected to rough handling, insert a sheet of heavy paper or cardboard over the explosive to keep it separated from the metal fragments to be added later.

4. Fill the remaining three fourths of the container with metal fragments or scraps. See Figure 45. Rocks, glass or other brittle materials are not satisfactory since they will be pulverized by the explosion.

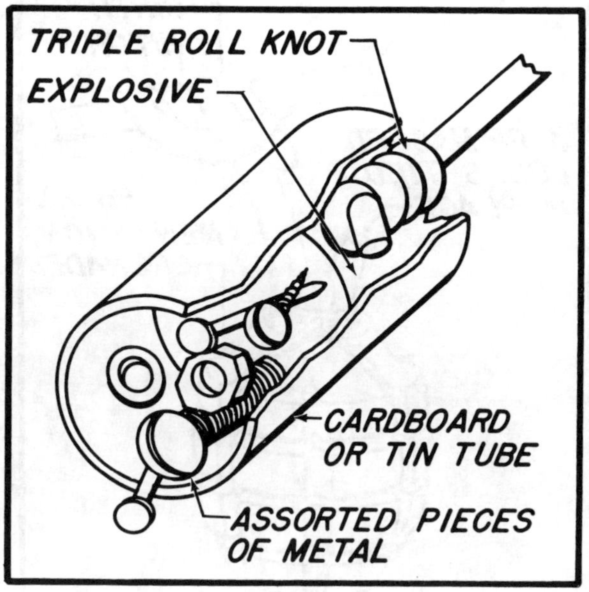

Fig. 45 — Scatter Charge

5. Fasten a cover of cardboard or cloth over the charge to keep the metal fragments from falling out of the container.

For ambush use, the charge should be securely fastened into position above ground, and aimed down the trail or road along which the column of troops or vehicle convoy is expected to appear. It should, of course, be camouflaged. More than one charge may be used and fired simultaneously by using primacord hookup.

Similar charges having much the same effect may be made by using coiled chain (automobile tire chain size), short lengths of barbed wire, or serrated or grooved metal plates which break into fragments when the charge is exploded.

Frag Grenade Ambush

Fragmentation hand grenades dispersed along a trail or road and camouflaged provide an excellent ambush technique. With a primacord system any number can be fired simultaneously.

MATERIALS: String, wire or tape, fragmentation grenades, primacord, blasting cap.

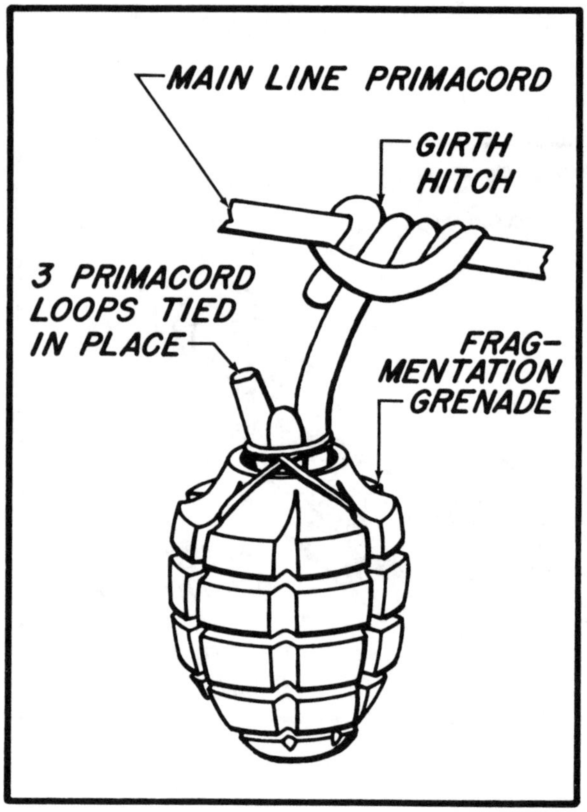

Fig. 46 — Frag Grenade Charge

PREPARATION:

1. Spread the points of the cotter pin safety to prevent accidental dislodging and then unscrew the fuse assembly of the frag grenade with pliers. The fuse may be saved for use as an improvised firing device or the blasting cap may be removed for other uses (see section on improvised detonators).

2. Cut a piece of primacord of whatever length needed to provide a branch line to a primacord main line. Usually about 3 or 4 feet is sufficient.

3. Fold one end of the primacord back on itself two or three times and wedge the folded end tightly into the empty fuse well of the grenade. Fasten the primacord securely to the grenade with wire, string or tape. See Figure 46.

4. Repeat the process for as many additional charges as are needed.

At the ambush site the grenades should be hung in bushes or tied to tree branches alongside the trail or road. If the grenades are placed on the ground (generally not the most effective placement) they should be upright, not lying on the side. Fasten the branch lines from the grenades to a main primacord line with a girth hitch or other tight knot. Attach a blasting cap and suitable firing device to the primacord main line. Camouflage the whole system.

For a heavy ambush the frag grenades may be interspersed with the scatter charge previously described, mortar shells, or with other improvised fragmentation charges.

Improvised Frag Grenade

Although more powerful than military fragmentation grenades, the improvised version described below is not as easily armed and thrown. It is advisable, therefore, to thoroughly test the time fuse used and perhaps to practice with a dummy model of the grenade before attempting to use a live one operationally.

MATERIALS: Metal fragments (nails, washers, bolts, nuts, screws, etc.), rags, string or tape, time fuse, nonelectric blasting cap, 1/2 pound of any high explosive.

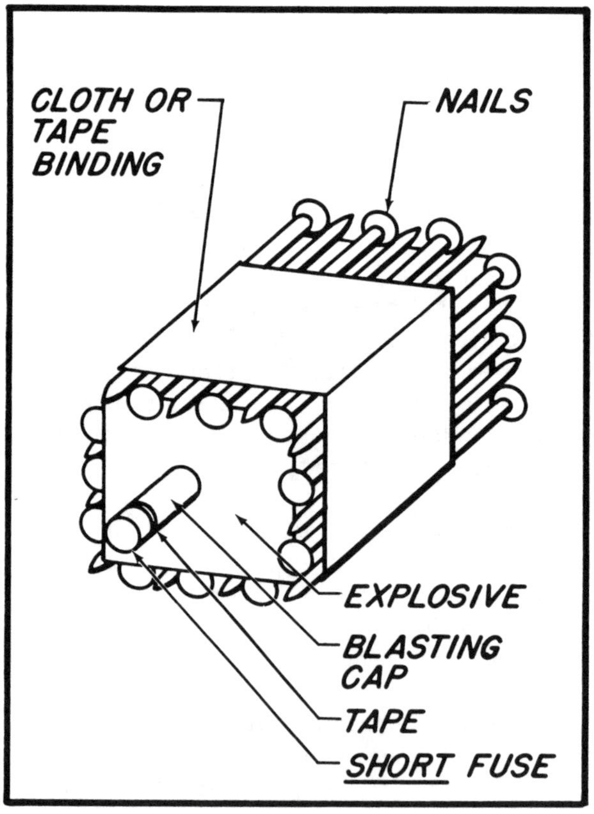

Fig. 47 — Improvised Frag Grenade

PREPARATION:

1. Surround a half pound lump or block of high explosive with metal fragments, leaving a small area clear for the insertion of a blasting cap. Nails may be neatly taped or tied to a block of explosive, as illustrated in Figure 47, or the metal fragments may be placed at random around a lump of explosive and the charge tightly wrapped in a rag or sock. It is important, however, that the charge be comfortable to hold and a handy size to throw.

2. Insert a piece of time fuse as far as it will go into a blasting cap. Mark the fuse about 1/2 inch from the opening of the blasting cap, remove it and cut it off at this mark.

3. Test the burning time of several pieces of fuse of the SAME LENGTH, cut from the SAME ROLL. By counting slowly and steadily, as "one thousand, two thousand, etc.," from the moment the fuse is lit to the time the spark is observed at the other end, a uniform delay of 6 to 7 seconds should be observed. If identical lengths of identical fuse do not burn uniformly, that fuse should NOT be used for this technique.

4. When a safe delay is reliably determined, prepare a firing system for the grenade by double crimping a blasting cap to the selected length of fuse. Wrap a short piece of tape around the joint between the cap and the fuse to prevent any spark from the lighting mechanism (match or fuse lighter) from bypassing the fuse into the blasting cap. Insert the cap into the charge and fasten it tightly in place.

This charge is most effective when an air burst is achieved directly over the target. The grenade should therefore be thrown with only 2 or 3 seconds of delay remaining. This is why it is necessary that the time fuse be fully tested in advance and that the user experiment until he can confidently count out the necessary delay after lighting the fuse.

As with any frag grenade, the user must

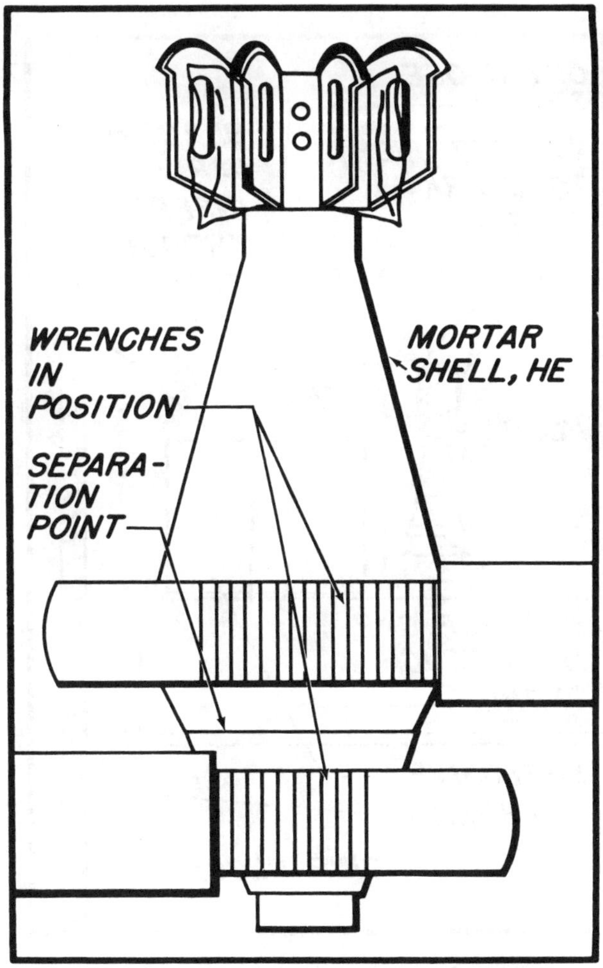

Fig. 48 — Removing Mortar Shell Fuse

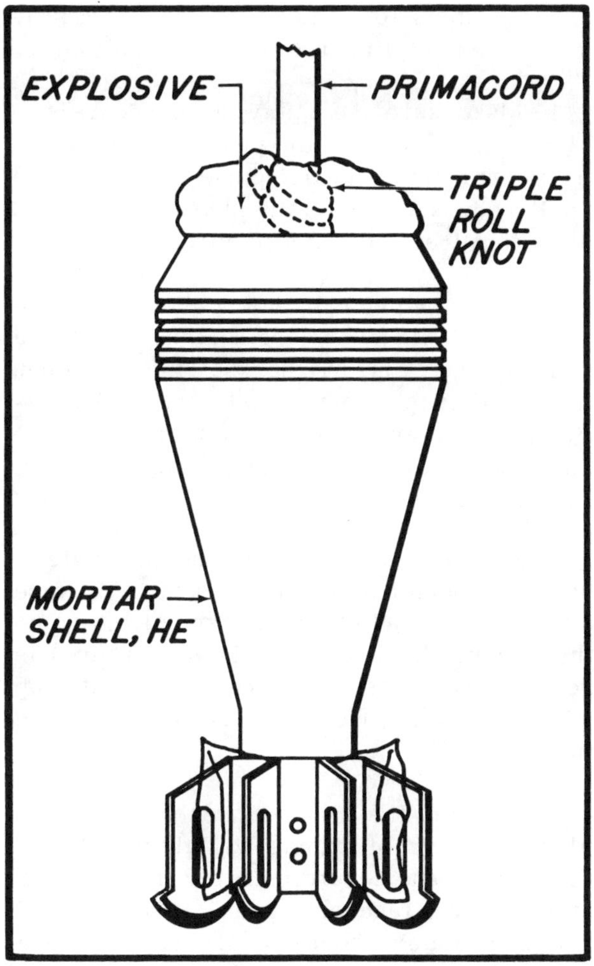

Fig. 49 — Priming Mortar Shell

protect himself from flying fragments. A prone position in a ditch, or shelter behind a tree, stump, or boulder will be sufficient protection.

Mortar Shell Ambush

Mortar shells can be fired as placed ambush charges in the same way as hand grenades. Being more powerful than grenades, they provide an even more lethal antipersonnel or antivehicular technique.

MATERIALS: String, wire or tape, mortar shell, primacord, blasting cap, plastic explosive (optional).

PREPARATION:

Many types of mortar shells are packed unassembled in separate containers. In such cases one need only fill the fuse well with plastic explosive and prime it with a primacord triple roll knot. Others, including U.S. shells, are packed fully assembled. Following are the steps to be taken to prepare a fully assembled shell for use as an ambush charge.

1. Make sure the safety pin is securely in place, then unscrew the fuse assembly from the mortar shell by using one wrench on the shell body and one on the fuse -- working in opposite directions. See Figure 48.

2. Tie a triple roll knot in the end of primacord of whatever length is needed for a branch line. Insert the knot into the open fuse well of the mortar shell, and pack plastic explosive around it. Fasten the explosive and knot in place with tape, wire, or string. See Figure 49.

3. Another way of priming the charge is to use the booster which is present in the fuse assembly instead of plastic explosive. To do this, leave the wrench on the fuse after it is removed from the shell, grip the booster cup with pliers and unscrew the booster from the fuse.

4. Replace the booster in the fuse well, place one turn of primacord tightly on top of the booster and secure it firmly in place with string, wire, tape or rags. The booster and primacord also should be securely fastened to the shell.

Any number of mortar shells may be prepared either with the plastic explosive technique or by using the booster and attached by their primacord branch lines to a main primacord line for simultaneous firing. Or they may be interspersed along a road or trail with frag grenades or other improvised charges.

Vehicle Booby Trap

Here's a safe and quick way of booby trapping an automobile without wiring into the ignition switch.

MATERIALS: Electric blasting cap, explosive.

PREPARATION:

1. Make a hole through a 1/2 pound or larger explosive charge (TNT, dynamite, plastic) so a blasting cap can be inserted all the way through the charge.

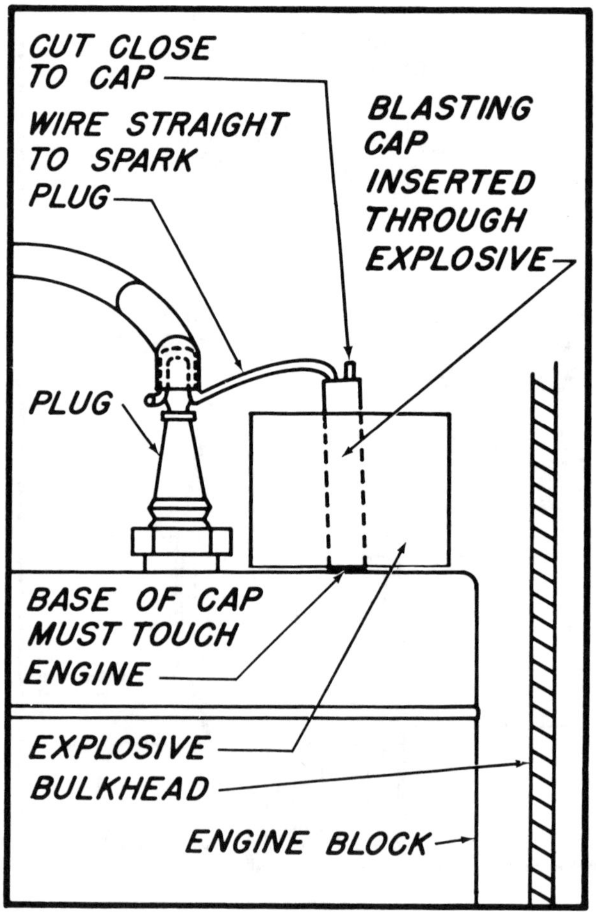

Fig. 50 — Vehicle Booby Trap

2. Snip off one of the two blasting cap wires up close to the cap and discard it.

3. Place the charge securely on the engine block of the target vehicle, as close as possible to the bulkhead of the passenger compartment.

4. Insert the blasting cap through the hole in the explosive so that the tip of the cap touches (is grounded against) the engine block. Except for the tip, the explosive end of the cap should be surrounded by explosive.

5. Stretch the remaining cap wire to the nearest spark plug. Cut off any excess wire, strip the end of the remaining portion and attach it to the top of the spark plug. See Figure 50.

6. Make sure that the cap wire is not near or touching any metal except the spark plug, otherwise there might be an accidental grounding at that point, resulting in a misfire.

The safety feature of this technique lies in the fact that the charge will not be activated until the engine is started, sending a spark through the plug. The ignition switch technique, on the other hand, can be dangerous unless the operator has a circuit tester or is thoroughly familiar with the ignition wiring.

THE 3.5 ROCKET

General

The U.S. 3.5" high explosive antitank rocket (or similar rockets used by other modern military forces) can be cannibalized or altered and used in a variety of improvised techniques. It can be placed as a shaped charge or land mine or it can be launched either electrically or nonelectrically without use of the rocket launcher. Such improvised firing offers the advantages of rigging the rocket as a booby trap or, at least, permitting the operator to be well out of the area before the rocket fires.

Placed Charges

MATERIALS: 3.5" rocket, safety fuse, primacord, blasting cap, plastic explosive.

PREPARATION:

1. Leaving the safety band in place, unscrew the warhead from the rocket with pipe wrenches, or wrench and a vise, as illustrated in Figures 51 and 52.

2. The detonator booster may remain connected to the warhead or it may unscrew as part of the fuse section, depending upon the tightness of the respective threads. If the booster remains connected to the warhead, handle it with care. The exposed red dot on the booster is a stab-sensitive detonator. Cover this with tape or other protection until ready to prime the charge.

3. If the booster remains in the warhead, the charge can be detonated by attaching a blasting cap or primacord across the red dot.

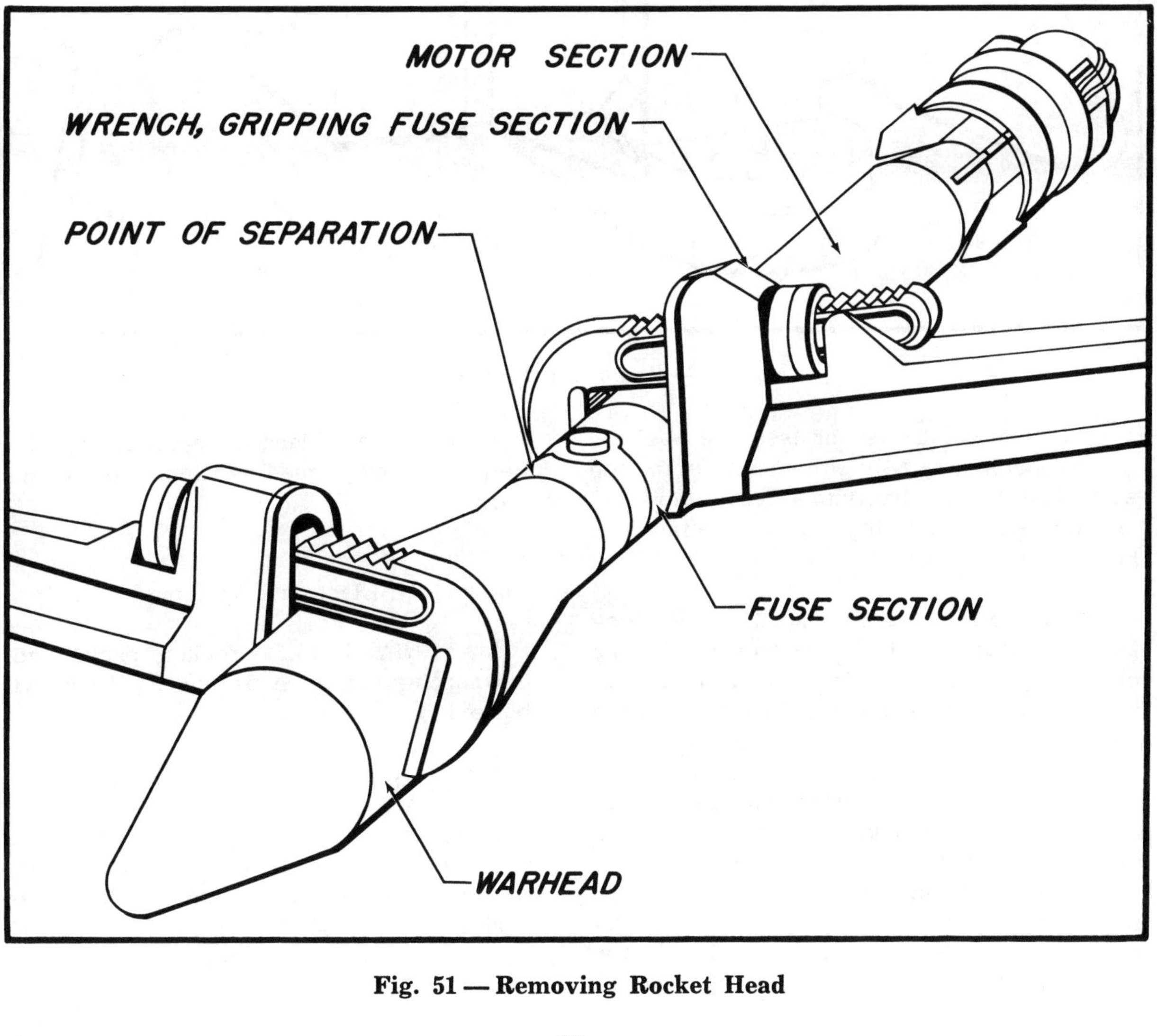

Fig. 51 — Removing Rocket Head

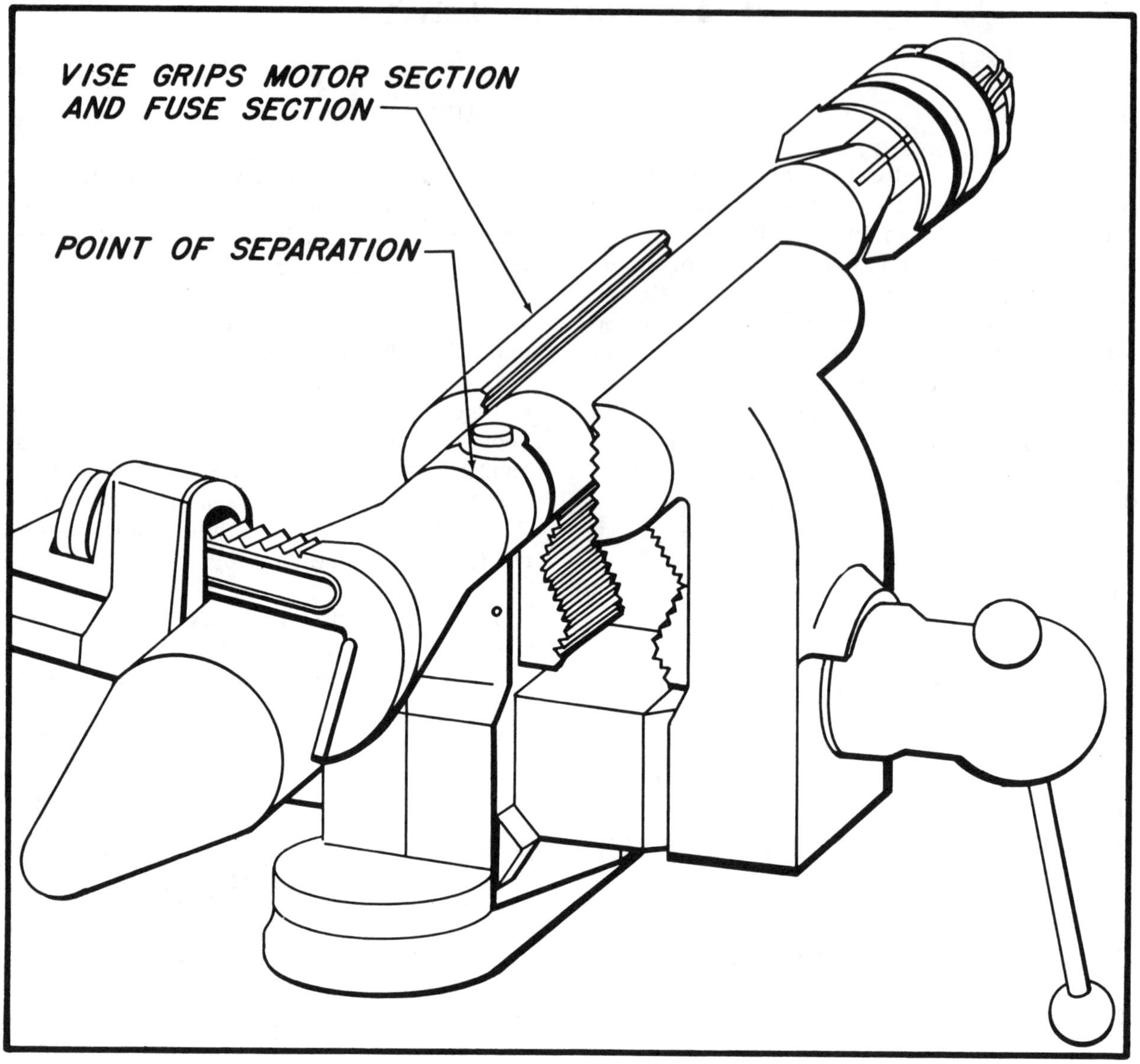

Fig. 52 — Removing Rocket Head

4. If the booster is not used, the warhead may be primed by filling the cavity in the top with plastic explosive, into which is embedded and fastened a blasting cap or a primacord knot.

5. To use as a placed shaped charge, prepare a tripod of sticks taped or tied to the charge to hold it upright on the target as illustrated in Figure 53. The nose of the charge may be directly against the target, or a slight additional standoff can be provided, as illustrated. For awkward shapes such as generators or turbines, the charge can be placed against the side of the target by using a counterweight system. Approximately 15 inches penetration of mild steel can be obtained with a properly prepared 3.5 rocket head shaped charge.

6. To use as a land mine, the charge should be placed and primed as illustrated in Figure 54.

Improvised Electric Launching

MATERIALS: 3.5" rocket, packing tube, downspout, soil pipe or boards, firing wire, batteries.

PREPARATION:

1. First find the two wires in the rocket tail section which disappear through a plastic plug into the rocket motor. (In the current model of the 3.5, these are the two wires covered with clear plastic.)

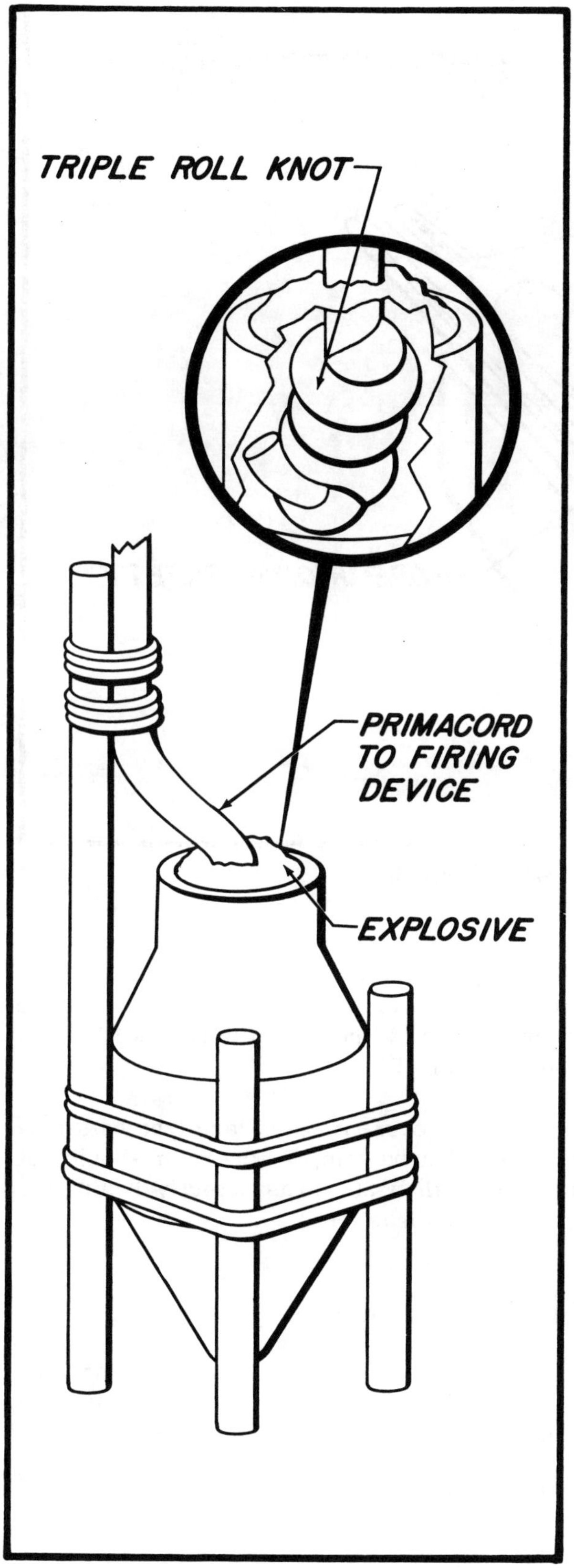

Fig. 53 — Rocket Shaped Charge

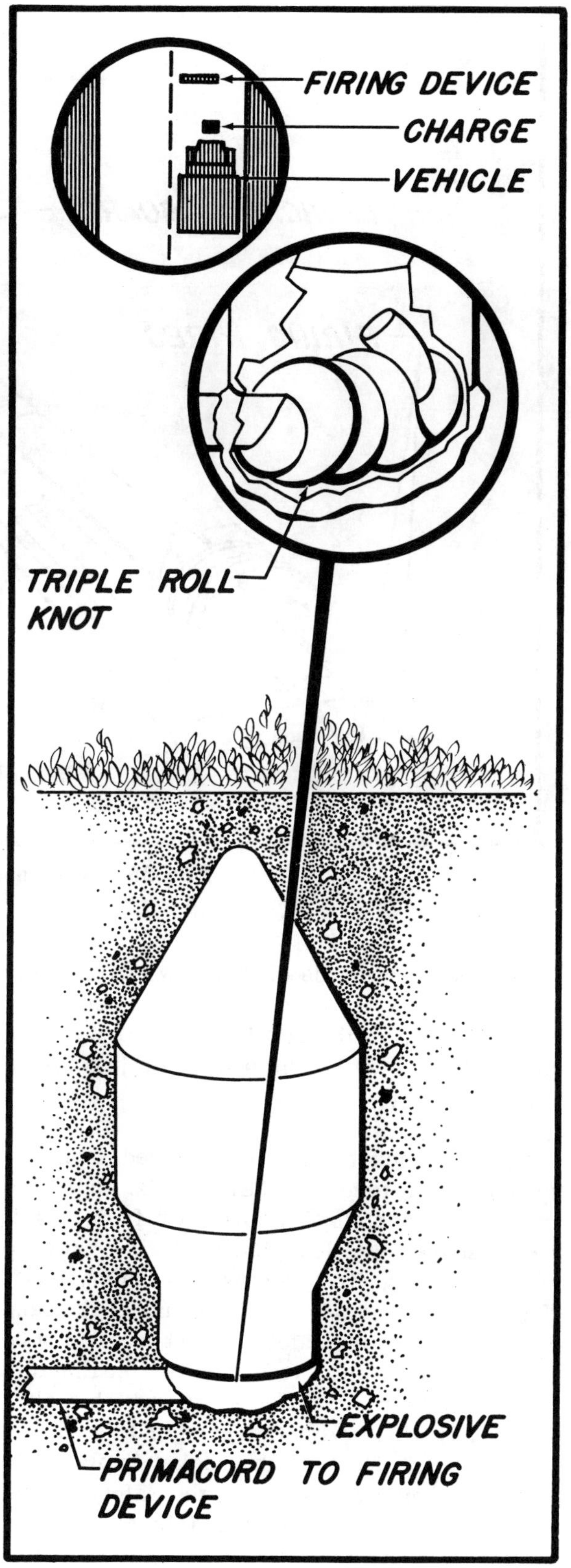

Fig. 54 — Rocket Land Mine

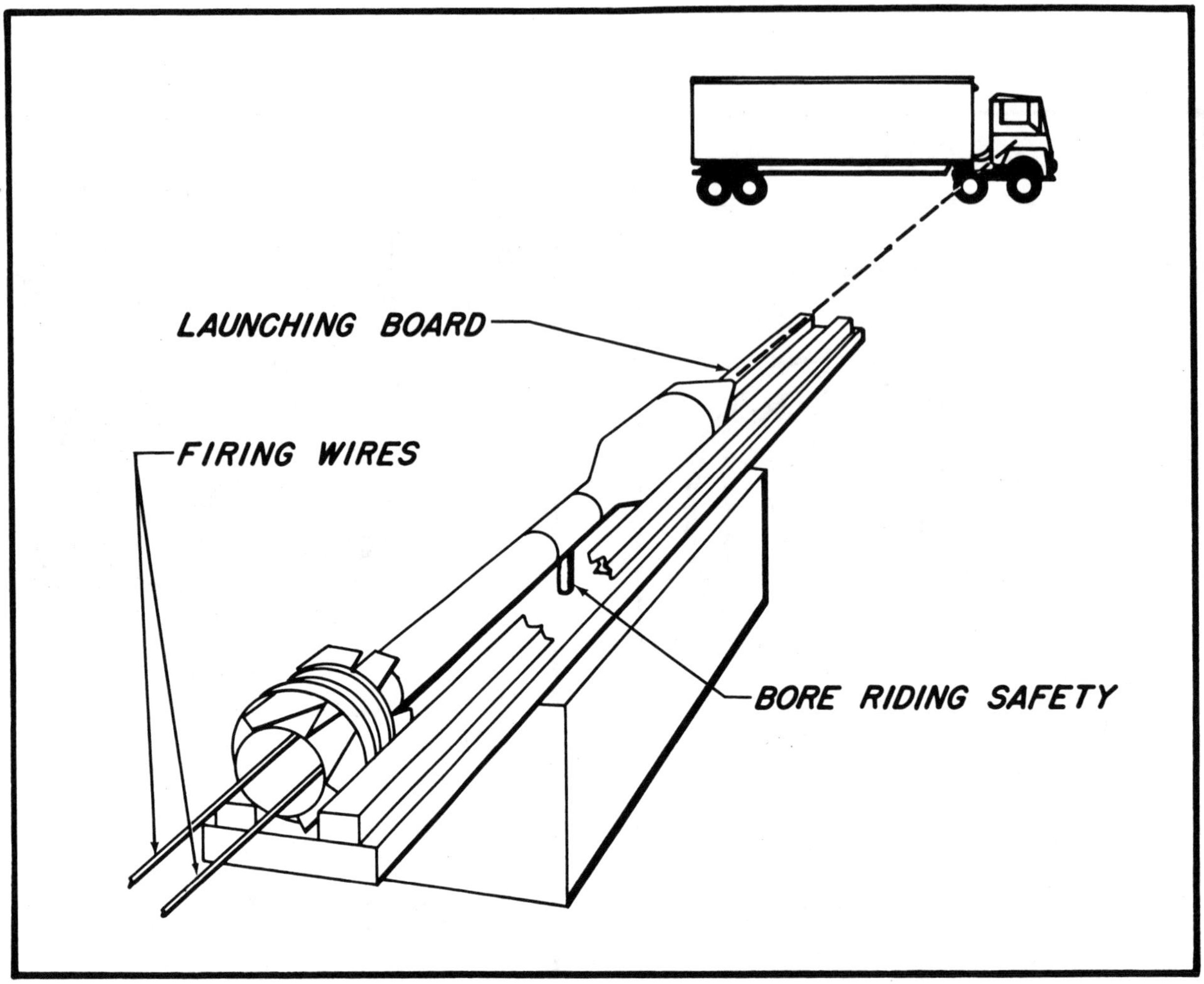

Fig. 55 — Improvised Rocket Launcher

2. Detach the ends of these two wires from any others to which connected and strip the insulation from the last half inch or so. The other wires may be snipped off or just folded out of the way.

3. The rocket may be launched from its own packing tube or other tube (as indicated under "materials," above), or from a flat or V-shaped wooden platform. To prepare for launching, first remove the safety band. The small spring-loaded button under the band is the bore-riding safety. The rocket must be placed in the improvised launcher so that this safety is depressed to a degree which makes it exactly flush with the head and the tail of the rocket. Thus, if the rocket is placed in a tube or on a flat surface, the safety should be on the bottom. If placed in a V-shaped launcher, the safety should be against one of the sides. See Figure 55 for proper placement on an improvised launcher.

4. After the rocket is placed on the launcher and sighted (of which more later), attach firing wires to the rocket wires as illustrated in Figure 56.

5. Apply current with flashlight batteries to fire. To be safe, the operator should not be closer than 50 yards directly behind the rocket when it is fired.

Nonelectric Launching

MATERIALS: 3.5" rocket, packing tube or other launching material as described above, safety fuse, safety matches.

PREPARATION:

1. To prepare the rocket for nonelectric launching, first remove the plastic plug (previously referred to) which blocks the motor

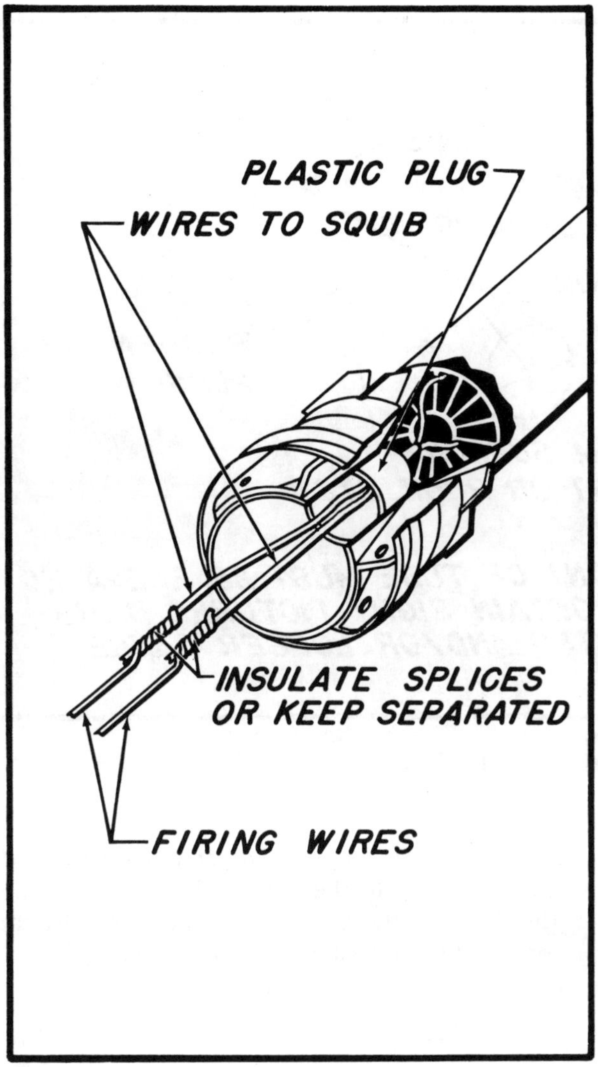

Fig. 56 — Electric Firing

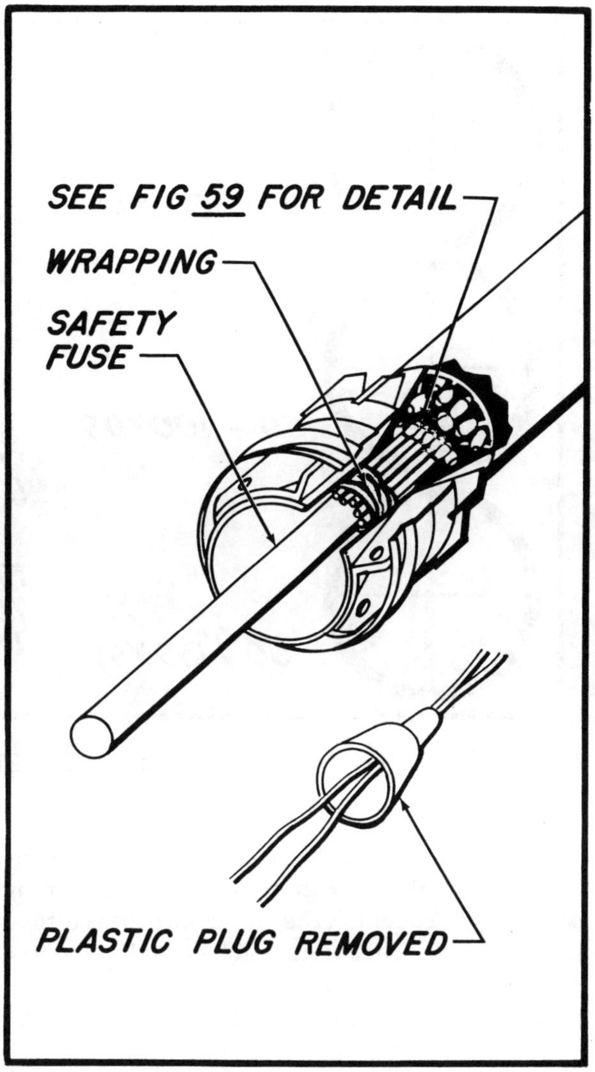

Fig. 57 — Nonelectric Firing

vent tube by prying loose with the pointed handle of crimpers.

2. Break off at least 20 safety match heads and drop them down into the cavity beyond where the plastic plug was removed.

3. Prepare a time fuse fire starter as described and illustrated in the following section on "improvised incendiaries." See Figure 59. Insert the fire starter among the match heads and wedge it securely in position with paper, tape, or parts of a match box. See Figure 57.

4. The rocket can now be fired from an improvised launcher by lighting the time fuse.

ALTERNATIVES: A spoonful of match head igniter (match head scrapings) can be used in place of the 20 match heads. A magnesium head time pencil, or a time pencil to which is attached a small sack or paper tube loosely filled with black powder, can be substituted for the match heads and the time fuse fire starter.

Ranges and Aiming

The minimum range for the 3.5 rocket is 15 feet, which is the distance required for the bore-riding safety to fly free and arm the rocket. The best point blank range is 15 feet to 100 feet.

Because removal of the plastic plug lowers the initial pressure, the range of the rocket when fired nonelectrically will be a little reduced and aiming should be a little higher.

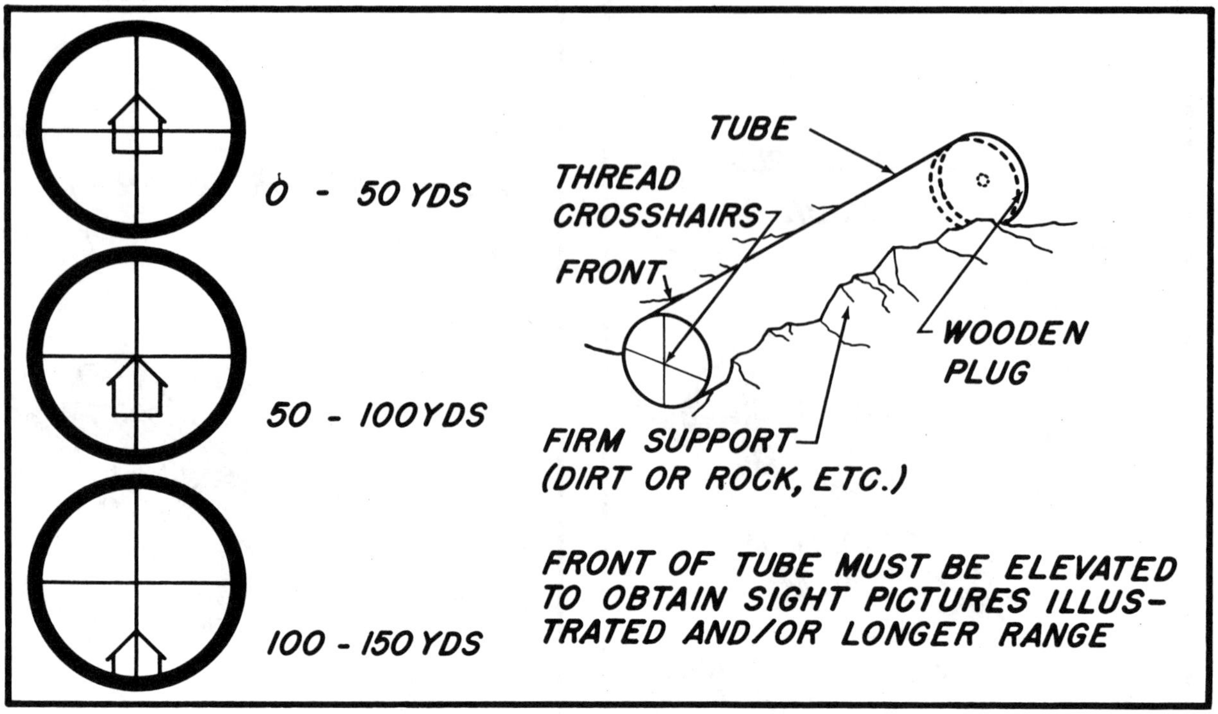

Fig. 58 — Sighting of Rocket

With practice a high degree of accuracy can be obtained with improvised launching of the 3.5 rocket. The inner cardboard tube in which the rocket is packed makes an excellent launcher. Remove the staples which hold the wooden plug in the end of the tube so the plug can be easily removed and reinserted. Attach crosshairs of thread or other material to the front of the tube, as illustrated in Figure 58, and sight in on the target through the peephole in the wooden plug. When the tube is sighted in on the target, wedge the tube firmly in that position, remove the plug and insert the rocket from the rear.

Figure 58 illustrates points of aim for different distances.

IMPROVISED INCENDIARIES

General

Good incendiaries can be improvised more easily than explosives and the materials are more easily obtained. On a pound for pound basis, incendiaries can do more damage than explosives against many type targets if properly used. There is a time lag, however, between the start of a fire and the destruction of the target. During this period the fire may be discovered and controlled or put out. An explosive once detonated has done its work.

Incendiaries are cheap and little training is needed for their preparation and use. Used in very carefully executed operations, the act of sabotage may be concealed in the ashes of an "accidental" fire.

Fires may be started quickly and have reasonable chance of success if the following few simple principles are observed:

1. See that there is plenty of air and fuel to feed the fire.

2. Use an incendiary that supplies a prolonged and persistent heat.

3. Start the fire low in the target structure and let it spread naturally upwards.

4. Use reflecting surfaces, such as corners, boxes, shelves, to concentrate the heat.

5. Use drafts to spread the fire rapidly -- near stairways, elevator shafts.

6. Protect the fire from discovery during the first few minutes by good concealment and timing.

In preparing improvised incendiaries observe basic rules of safety. Chemicals that must be powdered should be ground separately with clean tools and then mixed in the indicated proportions. Chemicals or mixtures should be kept tightly sealed in jars or cans to protect them from moisture. Damp materials will work poorly if at all.

Sulfuric acid, which is useful for chemical delays and to ignite incendiaries or explosive detonators, can be obtained by concentrating battery acid. This can be done by boiling off the water in the battery acid in a glass or porcelained pan until dense white fumes begin to appear. This operation should be done out of doors and the resulting concentrated acid should be handled carefully.

The paragraphs which follow will describe the preparation of several igniter (or "first fire") incendiary mixes, some basic incendiary mixes, and a thermate metal-destroying incendiary.

The subject of incendiaries has been treated much more exhaustively in other publications. The intent of this handbook is to provide only a few techniques.

Potassium Chlorate and Sugar Igniter

Chlorate-sugar is one of the best of the first fire or igniter mixes. It burns very rapidly, with a yellow-white flame, and generates sufficient heat to ignite all homemade incendiaries mentioned in this handbook.

MATERIALS: Potassium chlorate (preferred) or sodium chlorate, sugar.

PREPARATION:

1. Grind the chlorate separately in a clean, non-sparking (glass or wooden) bowl with a wooden pestle. The resulting granules should approximate those of ordinary table sugar.

2. Mix equal volumes of the granulated chlorate and sugar by placing both on a large sheet of paper and then lifting the corners alternately.

CAUTION: This mixture is extremely spark sensitive and must be handled accordingly.

3. Wrap 4 to 6 tablespoonfuls of the mixture in thin paper so as to form a tight packet.

Keep the mixture as dry as possible. If it is to be stored in a damp area before using, the packet may be coated with paraffin wax.

Chlorate-sugar is easily ignited by the flame of a match, the spit of a percussion cap or time fuse, the heat of an incandescent wire, or upon contact with concentrated sulfuric acid.

If ignited when under confinement, it will explode like gunpowder. If it is contained in a waxed packet, therefore, the latter should be punched through in several places before it is used with a basic incendiary and ignited.

Flake Aluminum-Sulfur Igniter

This simple igniter burns extremely hot and will ignite even the metal-destroying thermate described later on. The mixture itself can be lit by chlorate-sugar.

MATERIALS: Flake aluminum, finely powdered sulfur.

PREPARATION:

1. Mix 4 parts by volume of finely powdered sulfur with 1 part of aluminum powder.

2. Stir or shake in a closed container until the powders are thoroughly mixed and no lumps of sulfur remain.

To use, place several spoonfuls of the mixture on the material to be lit and add a spoonful of chlorate-sugar on top. Be sure the safety (time) fuse or other spark-producing delay system is placed so it will act upon the chlorate-sugar mixture first.

Homemade Black Powder Igniter

Black powder may be used for igniting napalm, flammable solvents in open containers, paper, loose rags, straw, excelsior and other tinder type materials. If it is not available already mixed, it can be prepared as follows:

MATERIALS: Potassium (or sodium) nitrate, powdered sulfur, powdered charcoal.

PREPARATION:

1. Into a clean, dry jar or can put 7 spoonfuls of potassium or sodium nitrate, 2 spoonfuls of powdered charcoal, and 1 spoonful of powdered sulfur. The ingredients must be at least as fine as granulated sugar. If they must be ground, GRIND EACH SEPARATELY. Never grind the mixed ingredients -- they may ignite or explode.

2. Cap the can or jar tightly and shake and tumble it until the ingredients are completely mixed.

The mixture will be effective for months if kept tightly sealed and dry. Sodium nitrate in particular has a tendency to absorb moisture.

To use the gunpowder, pile 2 or 3 spoonfuls on top of any solid incendiary material which is to be ignited. For igniting liquids in open containers, wrap 2 or 3 spoonfuls in a piece of paper and suspend it just above the liquid.

Gunpowder is best ignited by safety fuse. It burns very quickly and with a great deal of heat, so allow sufficient time delay for safe withdrawal from the vicinity.

Match Head Igniter

A good ignition material for incendiaries can be obtained from the heads of safety matches, which are available almost any place. The composition must be removed from the heads of many of them to get a sufficient quantity of igniter material. It will ignite napalm, was and sawdust, paper, and other flammables.

MATERIALS: Safety matches.

PREPARATION:

Remove the match head composition by scraping with a knife or crushing with pliers. Collect several spoonfuls of it and store in a moisture-tight container.

Put at least 2 spoonfuls on the material to be ignited. To ignite liquids, such as solvents or napalm, wrap several spoonfuls in a piece of paper and hang this just over the fluid, or place nearby. If fluids dampen the mixture it may not ignite.

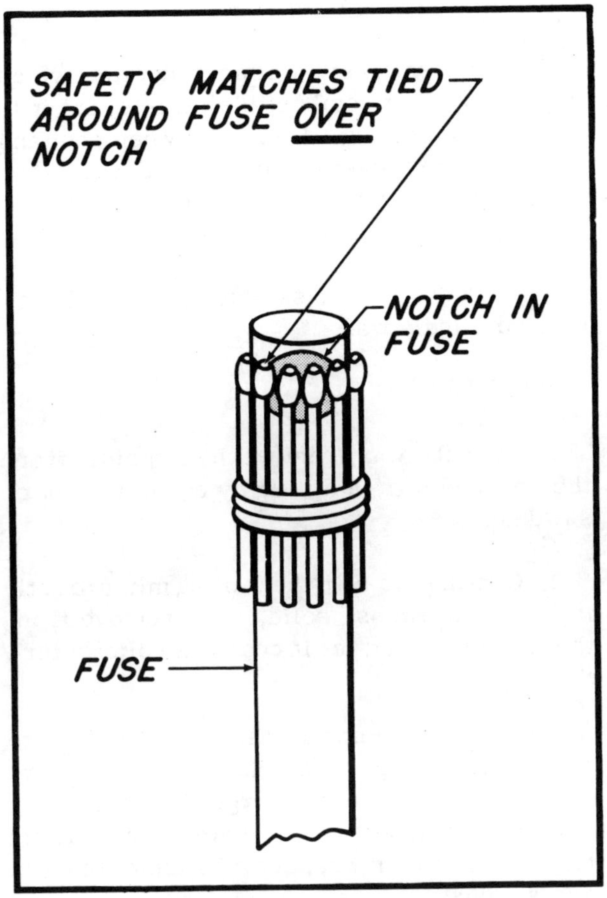

Fig. 59 — Time Fuse Fire Starter

Ignition can be by time fuse, firecracker fuse, a spark, or concentrated sulfuric acid.

Time Fuse Fire Starter

Several igniters or first fire mixes can be set off by a spark from time fuse. Others require a stronger flame. Time fuse, plus matches, can be combined to provide this more intense initial flame.

MATERIALS: Time (safety) fuse, safety matches, string or tape.

PREPARATION:

1. About 1/4 inch from the end of a piece of time fuse cut a notch with a sharp knife so that the powder train is exposed.

2. Around the fuse at this point tape or tie several matches so that their heads are in contact with each other and at least one match head is directly over the notch. See Figure 59.

When the fuse burns down, a spark from the notch ignites the one match head, which flares and ignites the others. This fire starter can be inserted into an igniter mix or used alone to light crumpled paper or excelsior. Another application, nonelectric firing of the 3.5" rocket, is described earlier.

Homemade Napalm

Napalm is the best incendiary to use against most flammable targets. It will readily ignite paper, straw, flammable solvents, or wooden structures.

MATERIALS: Gasoline or fuel oil, non-detergent soap (bar, flakes, or powder).

PREPARATION:

1. Use about equal parts of soap and oil. If bar soap is used, slice it into small chips. If both gasoline and fuel oil are available, use both in equal parts.

2. Heat the fuel in an open container, preferably one with a handle OUT OF DOORS. Try to avoid creating sparks or having a high open flame, but if the fuel should catch on fire, extinguish it by placing a board or piece of tin over the container.

3. Gasoline, in particular, will begin to bubble very quickly. When it does, remove

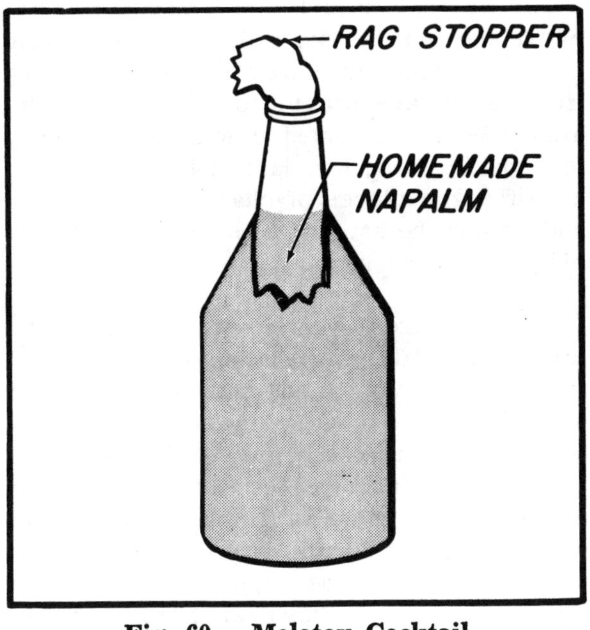

Fig. 60 — Molotov Cocktail

from the fire and gradually add the soap, stirring continuously, until the soap is completely dissolved and a thin pasty liquid results. If necessary return the mixture to the fire, but as a safety measure it is best not to stir while the container is on the fire.

4. When the desired consistency is reached allow the mixture to cool.

5. Napalm also can be mixed by a cold method, although it may take hours to thicken. This should be done by alternately adding very small amounts of soap chips or powder and gasoline or fuel oil and stirring until the mixture reaches a thin jelly-like consistency. It is best to start with about a cupful of soap, add part of a cup of solvent and stir that until smooth before gradually adding the remaining ingredients. Continuous stirring is not required. In fact, it is advisable just to let the mixture and the mixer rest from time to time and give the soap a chance to dissolve.

Napalm will keep well if stored in a tightly sealed container. It can be ignited with a match or any of the first fire mixtures described previously. The ignition packet should be placed adjacent to or just over the napalm, otherwise the petroleum may soak it and prevent its burning.

When napalm is used on easily ignitable materials (such as loosely piled paper, rags, or hay) it should be spread out so it will start a large area burning at once. Tightly baled paper or rags should be loosened first, because they do not burn well. If used directly against wooden structures or other large articles which are difficult to ignite, the napalm should be concentrated in sufficient quantity to provide a hot, long lasting blaze. If about a half dozen pieces of charcoal are put into and around the napalm the heat output is considerably increased.

Napalm makes an excellent "Molotov Cocktail." Just fill any glass bottle with a small neck with the napalm and cram a twisted strip of cloth into the mouth of the bottle as a stopper. See Figure 60. When ready to use, pull about 4 to 6 inches of the rag stopper out of the bottle. Light the napalm-soaked rag with a match and, when the rag is burning well, throw the bottle at the target. When the bottle breaks napalm splashes over the target and is ignited by the burning rag.

Paraffin-Sawdust Incendiary

Paraffin-sawdust is almost as effective as napalm against combustible targets, but it is slower in starting. It is solid when cool and thus is more easily carried and used than liquid napalm. In addition, it can be stored indefinitely without special care.

MATERIALS: Dry sawdust, paraffin, beeswax, or candle wax.

PREPARATION:

1. Melt the wax, remove the container from the fire and stir in a roughly equal amount of sawdust.

2. Continue to stir the cooling mixture until it becomes almost solid, then remove from the container and let it cool and solidify further.

Lumps of the mixture the size of a fist are easiest to manage. The chunks of incendiary may be carried to the target in a paper bag or other wrapper. Any igniter that will set fire to the paper wrapper will ignite the wax and sawdust.

A similar incendiary can be made by dipping sheets of newspaper into melted wax and allowing them to cool. These papers may then be crumpled up and used in the same manner as the paraffin-sawdust, although they will not burn as hot and persistently.

Sawdust, Moth Flakes, and Oil Incendiary

This incendiary is very good for use against all kinds of wooden structures, including heavy beams and timbers. It also works well on paper, rags, straw, excelsior, and other tinder type materials. It will start fires in open containers of flammable liquids, piles of coal, coke, or lumber, and on baled rags and paper. It is not effective against metal.

MATERIALS: Dry sawdust, moth flakes (naphthalene), fuel oil (kerosene or diesel oil).

PREPARATION:

1. Place equal parts of sawdust, moth flakes, and oil into a container and stir until the mixture is the consistency of mush.

2. Store it in any container that will retain the oil fumes.

An easy, effective way to use this mixture is to put about a quart of it in a paper bag and place the bag on the target material. The bag can be lit with a match and the mixture will ignite quite readily. It burns as well as napalm. If a longer delay time is required, use one of the igniter mixes described earlier along with time fuse or other delay device. The time fuse alone, however, will not ignite the incendiary mix.

Where very large wood beams are to be burned, an additional amount of the incendiary will be required. Two or three quarts is enough to destroy almost any target against which the technique would be effective.

For the greatest effect on wooden structures, the mixture should be in a pile, never spread out in a thin layer. It should be placed beneath the target material, if possible, so the flames will spread upward. In a packing box or room, a corner is a good place to start the fire.

Thermate Incendiary

Thermate is similar to commercial thermit, used in welding, except that it also contains an oxidizer, making it easier to ignite. Thermate will readily burn paper, rags, excelsior, straw, and other tinder type materials. However, its main use in sabotage operations is against motors, gears, lathes, or other metal targets -- to weld moving parts together, warp precision machined surfaces, and so on. Since it burns with a brief, almost explosive action, it is not recommended for burning wooden structures or other materials where persistent heat is required.

A good source of ready-made thermate is the U.S. military AN M-14 Incendiary Grenade. To remove the thermate, first pry out the fuse assembly with crimpers or other nonsparking implement. See Figure 61. The reddish-brown caked substance on top of the contents of the grenade is a first fire mixture and it is spark sensitive. This should be broken up and the grayish powder beneath, which is the thermate, can be poured out.

Fig. 61 — Defusing Thermate Grenade

Thermate also can be made from aluminum or magnesium powder and a chemical oxidizing agent, as described below:

MATERIALS: Aluminum filings, powder or flakes, or magnesium filings or powder, plus any one of the following chemicals: potassium nitrate, sodium nitrate, barium nitrate, potassium dichromate, sodium dichromate, or potassium permanganate. Although aluminum and magnesium are equally effective, thermate made from magnesium is easier to ignite. Flake aluminum, which is the extremely fine variety used in paints, is excellent. In any case, both the metal and chemical ingredients should be no coarser than granulated sugar.

PREPARATION:

1. Fill a quart size (or larger) container about 2/3 full of equal parts of the metal powder and the oxidizing agent.

2. Cover with a tight lid, then roll and tumble the container until the contents are completely mixed.

3. If FLAKE aluminum is the metal used, fill the container 1/2 full of the aluminum, then add oxidizing agent until the container is 3/4 full. Mix as described above.

Thermate in a sealed container can be stored for months. To use, put 1 or 2 pounds of the mixture in a paper bag and place it on the target in such a way that when it burns the red hot molten material will run down and attack the vital parts.

Chlorate-sugar and aluminum-sulfur igniters are best for setting off thermate, particularly if the thermate contains aluminum powder, which is more difficult to ignite.

Thermate also is used in the improvised dust initiator and the external POL charges described later.

DUST AND VAPOR EXPLOSIONS

General

A dust or vapor explosion sometimes happens accidentally in industries, causing great damage. Several readily available materials are explosive when mixed with air in the form of a dust or vapor, so such explosions can be induced as a sabotage technique. Among suitable materials are flour, cornstarch, powdered milk, cocoa, very fine aluminum or magnesium powder, gasoline, and fuel oil.

To cause an explosion with any of these materials, it is necessary to scatter them through the air as vapor or dust and then ignite them. A device that can be made to do this is called a dust initiator. It is a small charge of explosive mixed with incendiary material. When detonated within or under a container of the dust or vapor material it will break the container, scatter the material through the air and then ignite the particles to create a second explosion.

The damage is greatest in a completely enclosed space since the destructive effect is due to the pressure of expanding gas pushing out walls and collapsing the building. Buildings with many windows or other large openings may not be badly damaged.

Very frequently an incendiary effect follows the explosion, setting fire to the debris, especially if gasoline, oil, or metals are used as the secondary explosion material.

High humidity or damp or wet materials will weaken the explosive effect of a dust initiator, even causing a complete failure.

Although it is not possible to estimate precisely the amount of dust or vapor producing material needed for a given target, too much or too little will decrease the effect. Experience has shown that for each 1,000 cubic feet volume of the target building you should use about 1-1/4 pounds of flour, cornstarch, etc., or 4/5 gallon of gasoline or kerosene. The initiating charge itself (explosive and incendiary) should be between 1 to 2 pounds to scatter and ignite 25 pounds of dry dust or 5 gallons of liquid fuel.

Improvised Dust Initiator

The dust initiator is a small explosive/incendiary charge which can be used to initiate dust or vapor explosions in closed areas such as buildings, rooms, boxcars, or closed storage areas.

MATERIALS: Explosive, thermate, blasting cap, and secondary material: flour, cornstarch, powdered milk, cocoa, gasoline or other volatile fuel.

PREPARATION:

1. Mix equal parts of explosive and thermate incendiary. A powdered explosive is easiest to use, and TNT may be safely curshed with a wooden implement in a wooden container. Plastic explosive or dynamite also may be used, although in this case the explosive should be placed under the thermate so that the thermate will be directed upward. The thermate can be obtained from the AN M-14 incendiary grenade or homemade as

Fig. 62 — Dust Initiator

described in the section on "improvised incendiaries."

2. If a dust or powder is to be used for the secondary explosion, the explosive/incendiary initiator may be enclosed in any sort of a packet and placed within or under the material to be dispersed and ignited. See Figure 62. If the initiator is to be used in gasoline or fuel it will have to be enclosed in a fuel-tight bottle and sealed with some material impervious to petroleum solvents. An alternative is to put the initiator under the container of fuel rather than in it. In all cases, the initiator should be beneath or within the material to be dispersed rather than on top of it. The best placement is in the center of the target area and at least a foot above floor level. If the charge is placed in a basement, the pressures will be confined and directed upward, increasing the total damage.

The charge may be detonated either electrically or nonelectrically by a blasting cap placed within the explosive/incendiary initiator. Any one of several of the delay systems previously described may be employed.

POL CHARGES

Internal Flotation Charge

This charge combines explosive and incendiary material arranged in a container which will float on the surface when dropped into a large POL storage tank. The advantage of using an internal charge is that the charge need not be as large as one placed externally. Also, being inside the tank the charge is concealed and in less danger of discovery if a long term delay is desirable. On the other hand, an external charge is less complicated to make and easier to place since it doesn't require access to the maintenance and inspection hole in the roof of the tank.

MATERIALS: Fuel-tight container (approximately 2 gallon volume), plastic explosive, thermate, primacord, nonelectric blasting cap, delay device.

PREPARATION:

1. Cut a piece of primacord about 12 inches long. Tie a triple roll knot in one end and a girth hitch or other tight knot a few inches above the end knot.

2. Pack about 2 pounds of plastic explosive tightly into the bottom of the container and imbed the triple roll knot in it.

3. Add about the same quantity of thermate on top of the explosive. The thermate may be that obtained from the AN M-14 incendiary grenade or homemade, as described in the section on "improvised incendiaries." A coarse grain thermate (coffee ground size or larger) will work better in this technique than powdered thermate, and magnesium is preferable to aluminum.

4. Add another layer of plastic explosive equal to the bottom layer and be sure the girth hitch knot is imbedded in this layer. The sandwiched layers of explosive and thermate should occupy no more than the bottom two-thirds of the container, the rest being open air space to provide buoyancy of the charge (in fuel oil or gasoline, not water) prior to placement in the target. If additional buoyancy is needed, a block of wood can be added in the top of the container.

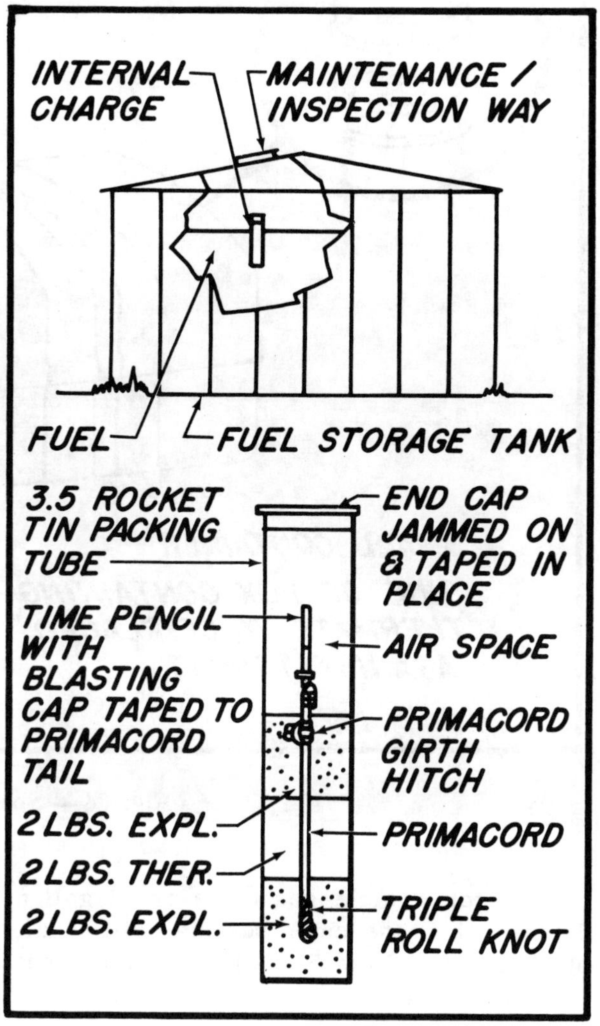

Fig. 63 — Internal Flotation Charge

Figure 63 shows the makeup of the internal flotation charge and the final preparations necessary for priming. To prime, attach the blasting cap with a time pencil or other delay device to the top end of the primacord, initiate the delay and seal in the whole charge and firing system with an airtight cap. The charge can then be dropped into the tank and should float upright until the delay time expires and the charge is fired.

External Charge

A plastic soap box, a cigar box or any such container can be used to fashion a simple charge for attacking fuel tanks from the outside. The soap box size charge is large

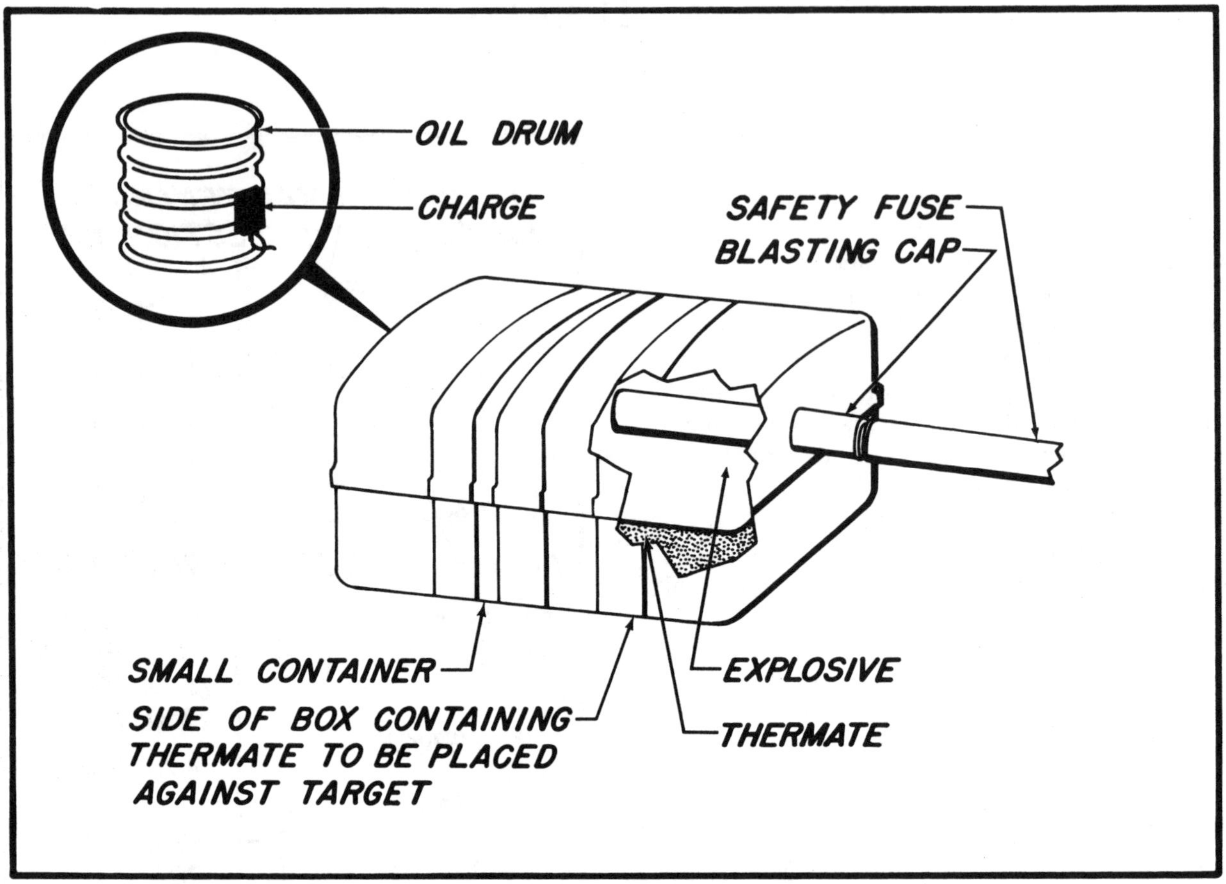

Fig. 64 — Soap Box Charge

enough for vehicle gas tanks or 55 gallon drums, while the cigar box size will handle storage tanks from 55 to 10,000 gallons capacity.

MATERIALS: Container, thermate, plastic explosive, nonelectric blasting cap, safety fuse, primacord (optional).

PREPARATION:

1. Fill the container half full of coarse thermate (obtained from the AN M-14 incendiary grenade or homemade).

2. On top of the thermate tightly pack an equal amount of plastic explosive, completely filling the container.

Prime the charge either with a triple roll primacord knot or a nonelectric blasting cap. Figures 64 and 65 illustrate properly made up charges.

The charge must always be placed so that the thermate side is against the target, otherwise the thermate would be blown away from the target when the charge explodes.

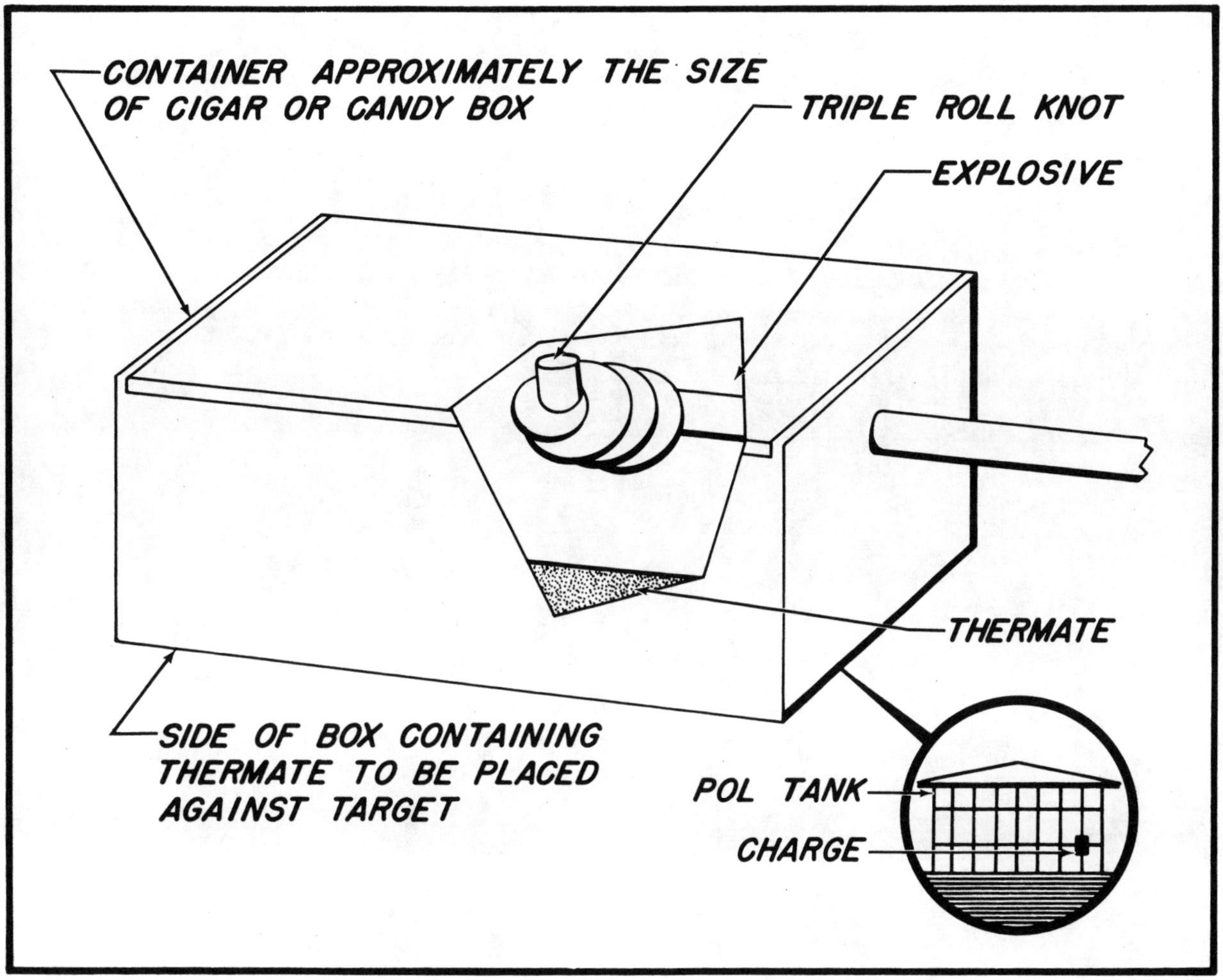

Fig. 65 — Cigar Box Charge

TARGET SUMMARY

Since many of the improvised devices and techniques described in the preceding sections of this handbook are applicable against a variety of targets, it seemed more feasible to organize the contents on a techniques rather than a target basis.

In summary, however, it might be helpful to review potential targets and list the techniques which could be used against them.

Notes

Notes

Notes

Notes

Notes

Other Books Available From Desert Publications

08/08

006	CIA Field Expedient Preparation of Black Powder	$19.95
221	Evaluation of Improvised Shaped Charges	$19.95
411	Homeopathic First Aid	$14.95
454	Survival Childbirth	$15.95
C-002	How To Open A Swiss Bank Account	$9.95
C-011	Defending Your Retreat	$9.95
C-020	Methods of Long Term Storage	$9.95
C-023	Federal Firearms Laws	$11.95
C-024	Select Fire Uzi Modification Manual	$13.95
C-025	How to Build Silencers	$6.95
C-027	Firearms Silencers Volume 1	$13.95
C-029	M1 Carbine Arsenal History	$7.95
C-032	The Silencer Cookbook	$9.95
C-038	How To Build A Beer Can Morter	$4.95
C-040	Criminal Use of False ID	$14.95
C-045	Improvised Weapons of American Underground	$14.95
C-048	Catalog of Military Suppliers	$18.95
C-056	Navy Seal Manual	$29.95
C-065	Poor Man's James Bond Vol 1	$34.95
C-113	Submachine Gun Designers Handbook	$19.95
C-114	AR-15, M16 and M16A1 5.56mm Rifles	$13.95
C-117	Ruger Carbine Cookbook	$13.95
C-120	Shootout II	$14.95
C-122	AK-47 Assault Rifle	$15.95
C-123	Combat Loads for Sniper Rifles	$18.95
C-125	Browning Hi-Power Pistols	$13.95
C-126	UZI Submachine Gun	$13.95
C-127	FullAuto Vol 1 Ar-15 Modification Manual	$11.95
C-128	Full Auto Vol 2 Uzi Mod Manual	$11.95
C-130	Full Auto Vol 4 Thompson SMG	$11.95
C-131	FullAuto Vol 5 M1 Carbine to M2	$11.95
C-135	M-14 Rifle, The	$12.95
C-136	Fighting Garand Manual	$13.95
C-137	Ranger Training & Operations	$19.95
C-138	Emergency War Surgery	$29.95
C-140	Emergency Medical Care/Disaster	$14.95
C-143	Survival Medicine	$15.95
C-150	Improvised Munitions Black Book Vol 1	$17.95
C-151	Improvised Munitions Black Book Vol 2	$17.95
C-152	Impro. Munitions Black Book Vol 3	$29.95
C-155	Fighting Back on the Job	$14.95
C-162	Colt .45 Auto Pistol Manual	$13.95
C-163	Survival Evasion & Escape	$19.95
C-165	USMC Sniping	$19.95
C-175	Beat the Box	$7.95
C-197	How to Make Disposable Silencers Vol. 1	$19.95
C-198	Expedient Hand Grenades	$21.95
C-199	U. S. Army Counterterrorism Training Manual	$19.95
C-203	Explosives and Propellants	$19.95
C-204	Improvised Munitions/Ammonium Nitrate	$19.95
C-205	Two Component High Explosives Mixtures	$19.95
C-207	CIA Improvised Sabotage Devices	$19.95
C-208	Spcial Forces Demolitions Training Handbook	$19.95
C-210	Guerilla Warfare	$17.95
C-211	FN-FAL Auto Rifles	$21.95
C-218	Professional Homemade Cherry Bomb	$15.95
C-219	Improvised Shaped Charges	$19.95
C-220	Improvised Explosives, Use In Detonation Devices	$19.95
C-227	Springfield '03 Rifle Manual	$15.95
C-229	Lock Picking Simplified	$11.95
C-230	How to Fit Keys by Impressioning	$9.95
C-231	Lockout -Techniques of Forced Entr.	$15.95
C-235	How to Open Handcuffs Without Keys	$13.95
C-244	Camouflage	$14.95
C-274	M14 and M14A1Rifles and Rifle Marksmanship	$23.95
C-284	Thompson Submachine Guns	$19.95
C-289	M1 Carbine Owners Manual	$13.95
C-290	Clandestine Ops Man/Central America	$14.95
C-324	How to Make Disposable Silencers Vol. 2	$19.95
C-337	Sten MKII SMG Construction Manual	$24.95
C-356	Map Reading and Land Navigation	$19.95
C-365	Survival Gunsmithing	$14.95
C-372	Poor Man's James Bond Vol 2	$34.95
C-386	Self-Defense Requires No Apology	$15.95
C-414	Select Fire 10/22	$15.95
C-415	Construction Secret Hiding Places	$15.95
C-416	Brown's Alcohol Motor Fuel Cookbook	$24.95
C-417	Canteen Cup Cookery	$9.95
C-418	Improvised Batteries/Detonation Devices	$19.95
C-421	Walther P-38 Pistol Manual	$13.95
C-422	P-08 Parabellum Luger Auto Pistol	$14.95
C-423	Sten Submachine Gun, The	$11.95
C-427	How to Train a Guard Dog	$16.95
C-429	The Anarchist Handbook Vol. 1	$14.95
C-430	Beretta - 9MM M9	$15.95
C-432	Mercenary Operations Manual	$14.95
C-439	Improvised Lock Picks	$17.95
C-443	How to Disappear Completely & Never be Found	$19.95
C-455	Modern Day Ninjutsu	$17.95
C-512	FullAuto Vol 8 M14A1 & Mini 14	$11.95
C-518	L.A.W. Rocket System	$14.95
C-528	With British Snipers, To the Reich	$29.95
C-539	Dental Specialist	$59.95
C-558	Food. Fur and Survival	$17.95
C-561	The Squeaky Wheel	$17.95
C-562	Survival Chemist Book	$17.95
C-564	Medical Specialist	$59.95
C-571	U. S. Marine Corps Scout/Sniper Training Manual	$27.95
C-610	Aunt Bessie's Wood Stove Cookbook	$9.95
C-622	Defensive Shotgun	$14.95
C-625	MAC-10 Cookbook	$14.95
C-627	Brown's Lawsuit Cookbook	$26.95
C-628	Can You Survive	$16.95
C-633	Hand to Hand Combat	$11.95
C-634	USMC Hand to Hand Combat	$9.95
C-635	Hand to Hand Combat by D'Eliscue	$11.95
C-636	Prisons Bloody Iron	$16.95
C-639	Police Karate	$17.95
C-646	Science of Revolutionary Warfare	$16.95
C-668	Survival Shooting for Women	$14.95
C-680	Infantry Scouting, Patrol. & Sniping	$16.95
C-694	CIA Field Expedient Incendiary Manual	$19.95
C-761	How to Lose Your X-Wife Forever	$24.95
C-763	Vigilante Handbook	$19.95
C-775	Alcohol Distillers Handbook	$27.95
C-776	How to Build a Junkyard Still	$15.95
C-777	Engineer Explosives of WWI	$19.95
C-794	US Marine Corps Essential Subjects	$19.95
C-815	US Marine Bayonet Training	$11.95
C-818	The Poisoner's Handbook	$29.95
C-823	Cold Weather Survival	$14.95
C-829	The Anachist Handbook Vol. 2	$14.95
C-840	Freedom Road	$21.95
C-889	Napoleon's Maxims of War	$11.95
C-890	Company Officers HB of Ger. Army	$19.95
C-891	German Infantry Weapons Vol 1	$22.95
C-896	Op. Man. 7.62mm M24 Sniper Weapon	$13.95
C-899	US Army Bayonet Training	$11.95
C-9002	German MG-34 Machinegun Manual	$19.95
C-9012	Firearm Silencers Vol 3	$18.95
C-9013	HK Assault Rifle Systems	$27.95
C-9047	SKS Type of Carbines, The	$17.95
C-9048	Rough Riders, The	$24.95
C-9052	Private Weaponeer, The	$17.95
C-9057	Keys To Understanding Tubular Locks	$11.95
C-9060	The Anarchist Handbook Vol. 3	$14.95
C-9080	Leadership Hanbook of Small Unit Ops	$14.95
C-9081	Dirty Fighting	$13.95
C-9082	Lasers & Night Vision Devices	$29.95
C-9083	Water Survival Training	$8.95
C-9084	Assorted Nasties	$29.95
C-9085	Ruger P-85 Family of Handguns	$19.95
C-9101	Guide to Germ Warfare	$17.95
C-9102	CIA Field Expedient Method for Explosives Preparation	$19.95
C-9119	Surviving Global Slavery	$16.95
C-9120	Military Knife Fighting	$14.95
C-9127	H&R Reising Submachine Gun Manual	$19.95
C-9131	Live to Spend It	$29.95
C-9134	Military Ground Rappelling Techniques	$15.95
C-9135	Smith & Wesson Autos	$19.95
C-9137	Caching Techniques of U.S. Army Special Forces	$14.95
C-9138	USMC Battle Skills Training Manual	$49.95
C-9139	Survival Armory	$27.95
C-9156	Concealed Carry Made Easy	$17.95
C-9164	The L'il M-1, The .30 Cal. M-1 Carbine	$17.95
C-9170	Urban Combat	$27.95
C-9178	Apocalypse Tomorrow	$17.95
C-9198	MP40 Machinegun Manual	$16.95
C-9199	Clear Your Record & Own a Gun	$24.95
C-9200	Sig's Handguns Manual	$19.95
C-9210	Sniper Training	$24.95
C-9212	Poor Man's Sniper Rifle	$19.95
C-9221	The Official Makarov Pistol Manual	$14.95
C-9224	The Butane Lighter Hand Grenade	$14.95
C-9228	Unarmed Against the Knife	$14.95
C-9229	Black Book of Booby Traps	$29.95
C-9239	Militia Battle Manual	$22.95
C-9247	Black Book of Revenge	$15.95
C-9253	Bankruptcy, Credit and You	$18.95
C-9255	The Sicilian Blade	$15.95
C-9259	How to Build Practical Firearm Silencers	$18.95
C-9262	Glock's Handguns	$19.95
C-9263	Heckler and Koch's Handguns	$19.95
C-9264	The Poor Man's Ray Gun	$11.95
C-9265	The Poor Man's R. P. G.	$24.95
C-9269	David's Tool Kit	$27.95
C-9279	Credit Improvement & Protection Handbook	$19.95
C-9281	Build Your Own AR-15	$19.95
C-9282	How to Live Safely in a Dangerous World	$19.95
C-9293	Bllack Book of Arson	$24.95
C-9322	How to Build Flash/Stun Grenades	$19.95
C-9342	Entrapment	$19.95
C-9359	The FN-FAL Rifle Et Al	$21.95
C-9367	The Mental Edge, Revised	$17.95
C-9373	Survival Bible by Duncan Long	$59.95
C-9405	The Internet Predator	$17.95
C-9406	The Criminal Defendants Bible	$49.95
C-9430	Protect Your Assets	$24.95
C-9432	How To Build Military Grade Suppressors	$21.95
C-9441	Mossberg Shotguns	$24.95
C-9478	Poor Man's Primer Manual, The	$24.95
C-9546	Agents HB of Black Bag Ops.	$14.95
C-9556	Improvised Rocket Motors	$19.95
C-9612	Homestead Carpentry	$14.95
C-9633	Getaway	$13.95

PRICES SUBJECT TO CHANGE WITHOUT NOTICE

800-852-4445

Desert Publications

215 S. Washington Ave. Dept. - 209
El Dorado, AR 71730 USA
info@deltapress.com

Shipping & handling
1 item $7.95 - 2 or more $12.95